Promoting Geography in Schools

Promoting Geography in Schools

Edited by
Ashley Kent

THE GEOGRAPHICAL ASSOCIATION

© the Geographical Association, 1999

This book is copyright under the Berne Convention. All rights are reserved. Apart from any fair dealing for the purpose of private study, research, criticism or review, as permitted under the Copyright, Designs and Patents Act 1988, no part of this publication may be reproduced, stored in a retrieval system, or transmitted in any form or by any means, electronic, electrical, chemical, mechanical, optical, photocopying, recording or otherwise, without the prior written permission of the copyright owner. Enquiries should be addressed to the Geographical Association. As a benefit of membership, the Association allows its members to reproduce material for their own internal school/departmental use, provided that the copyright is held by the GA.

ISBN 1 899085 57 2
First published 1999
Impression number   10 9 8 7 6 5 4 3 2 1
Year   2002   2001   2000   1999

Published by the Geographical Association, 160 Solly Street, Sheffield S1 4BF. The Geographical Association is a registered charity: no 313129.

The Publications Officer of the GA would be happy to hear from other potential authors who have ideas for geography books. You may contact the Officer via the GA at the address above. The views expressed in this publication are those of the authors and do not necessarily represent those of the Geographical Association.

Illustrations by Dave Howarth
Designed by Ledgard Jepson
Printed and bound in England by J W Northend Ltd.

Acknowledgements

We would like to acknowledge the work of the following people - experienced, effective and enthusiastic geographers - who have contributed to this book:

- Tony Dodsworth is Head of Humanities at Pope Pius X RC Comprehensive School, Wath-upon-Dearne, Rotherham
- Peter Fry is Deputy Headteacher, Edgbarrow School, Crowthorne, Berkshire
- Keith Grimwade is General Advisor, Geography, for Cambridgeshire
- Huw Jackson is Head of Geography at Bexleyheath School, Kent
- Ashley Kent is Head of Education, Environment and Economy at the Institute of Education, University of London
- Alan Marvell is Head of Geography, New College, Swindon
- Garrett Nagle is Head of Geography at St Edward's School, Oxford
- Tim Price-Walker teaches geography at Downe House School, Thatcham, Berkshire
- Patrick Talbot is Head of Careers and a member of the Geography Department at Hampton School, Middlesex

The Geographical Association would also like to thank Judith Mansell (Head of Geography, Bishop's Hatfield Girl's School, Hertfordshire) for valuable additions. Finally, the Association is grateful to the following organisations for allowing the reproduction of material: the Southern Examining Board for the statistics on GCSE Travel & Tourism results, the Royal Geographical Society (with the Institute for British Geographers) for the information on geography graduates and the Careers and Occupational Information Centre (COIC) for Figures 1 and 2 in Chapter 5.

Contents

Introduction — page **7**

Chapter 1: **Campaigning for geography** — page **9**
Ashley Kent

Chapter 2: **Living geography** — page **13**
Garrett Nagle

Chapter 3: **Tailoring your approach to your audience** — page **23**
Tim Price-Walker with Peter Fry, Keith Grimwade and Alan Marvell

Chapter 4: **Promoting geography to your school's senior managment team** — page **36**
Keith Grimwade

Chapter 5: **A world of opportunities** — page **39**
Patrick Talbot with Alan Marvell

Chapter 6: **Extra-curricular activities in geography clubs and societies** — page **51**
Tony Dodsworth

Chapter 7: **Showing some initiative: a case study** — page **59**
Huw Jackson

Bibliography — page **63**

Organisations and addresses — page **64**

Promoting Geography in Schools

Introduction

Many geography teachers increasingly find themselves having to 'market' their subject to a variety of audiences, including senior management teams and other colleagues, current and prospective students, parents or members of the local community. Terminology previously associated with commercial marketing is therefore starting to make itself heard in the classroom.

What is geography's USP (unique selling point)? Where does the competition lie? How, and when, should different audiences be targeted? How much precious time and budget should busy geography teachers devote to 'marketing' their subject?

It is against this background - and to fulfil that part of the GA's mission defined as 'strengthening the position of geography' that the GA has adopted a three tiered publishing response to meet the needs of those members wishing to strengthen the position of geography in their schools. This will comprise:

- *Promoting Geography in Schools.* In this book classroom teachers offer ideas for activities and events which will emphasise to students and parents the importance of geography both as a subject in its own right, and as a gateway to many varied and interesting careers.
- *Going Places: A geography careers resource pack.* This is highly complementary to *Promoting Geography in Schools*, but takes a more practical approach. It contains a teachers' book which offers the basics of careers guidance and a detailed bibliography and reference section, and templates for handouts, leaflets, colour posters.
- *Teaching Geography.* The GA's journal for secondary geography teachers will regularly feature case studies, up-to-date statistics and ideas for promoting geography to supplement and refresh ideas presented in *Promoting Geography in Schools* and *Going Places: A geography careers resource pack*.

Promoting Geography in Schools is based on its highly successful predecessor, *Selling Geography* (Kent, 1990), which both identified the need to 'sell' geography and provided a number of mechanisms for doing it.

The objectives of the book are:
- to inform parents, students and others of what school geography is about and why it is important to study, and thereby
- to increase the numbers studying geography in schools

All contributors are practising classroom teachers or advisers who draw on their own experiences to provide innovative ideas for promoting geography in a range of contexts and to different audiences. These include:

- Ideas for emphasising the relevance and topicality of geography in the classroom, thus making geography lessons more relevant and interesting.
- Case studies which demonstrate how, by adopting well thought out 'marketing plans', demand for the subject can be sustained and increased.
- Checklists and action plans to help busy teachers maximise the opportunities presented by open evenings.
- Ideas for making the most of geography's career potential.
- Practical implications and creative opportunities of setting up a geography club to foster and encourage enthusiasm for the subject.

Promoting Geography in Schools therefore offers a comprehensive range of tried and tested ideas for promoting geography to secondary students and their parents.

Further opportunities for marketing the subject are provided by Geography Action Week, organised annually by the Geographical Association to celebrate and promote geography in schools and beyond.

Meanwhile the Geographical Association continues to represent the subject at the highest levels in order to safeguard and extend geography's contribution to education.

Promoting Geography in Schools

Chapter 1:
Campaigning for geography
Ashley Kent

Promoting geography in schools and universities is part of a wider campaign for geography. You can sell geography to prospective students in the classroom, to their parents at open days/evenings in schools and at introductory visits to universities. You can promote geography by maintaining a dialogue with colleagues, with local and with national employers. Membership of a national subject association (e.g. the Geographical Association) ensures that your subject is represented nationally to policy makers and opinion formers. National and regional conferences and press/journal articles offer you the opportunity to keep up-to-date with the most recent concerns in geography. For instance, at a national event 'Education for Life: a dialogue about the contribution of geography', organised jointly by the RGS (with IBG in June 1997) and the GA, Michael Bichard (Department for Education and Employment) posed specific challenges by asking:

- 'What is geography and what is it for?'
- 'How should geography fit into the whole curriculum and what is its impact on other subjects?'
- 'What are the special arguments for geography?'

To which Dr Nick Tate (of the Qualifications and Curriculum Authority) responded:

'One of the issues for subjects such as geography is the extent to which they can contribute to learning how to learn, learning why it is important to learn, learning to work with others, learning how to internalise the shared values of our society, learning to live in a community ... as well as to literacy and numeracy across the curriculum.'

He argued that geography's potential contribution 'needs to be made explicit and to become part of the subject's *raison d'être* in the minds of all who teach it and learn it'. You must remember that geography is for all, irrespective of ability, gender, ethnicity and social circumstances. Plan your curriculum accordingly and *promote it* to all (see Barrett, 1996 and Swift, 1996). This book helps meet these challenges by showing how professional geographers use a range of strategies to present a modern subject through:

- Selling geography to students through interesting and topical studies (see Chapter 2 and e.g. Conolly, 1997) and use the photocopiable leaflets from the GA's new careers pack (Palôt, 1999).

- Selling geography to parents at open evenings (see Chapter 3).

- Challenging stereotypes - by presenting examples of females in exciting careers related to geography.

- Obtaining positive media coverage for the subject in your school or local education authority.

- Influencing the headteacher and senior management team by constantly updating them with your department's activities, and lobbying for their support when it is needed for departmental developments or capital expenditure (Chapter 4).

- Promoting geography in schools (see Chapters 1, 2, 3 and 7 and e.g. Jackson, 1996). You can also do so by taking part in the annual Geography Action Week organised by the Geographical Association and setting up your own school geography club (see Chapter 6).

- Maintaining contact with local and national employers and with former students (see 'Pen portraits' below and Chapter 5) as evidence of the usefulness of geography in employment. Keep copies of careers information leaflets/packs (see RGS (with IBG)/GA, 1998; GA, 1998 and Palôt, 1999).

- Involving yourself with the dialogue between policy makers and opinion formers from both the public and private sectors by reading and/or contributing articles to the press/journals (e.g. Kent, 1996).

- Keeping copies of videos (e.g. Scottish Association of Geography Teachers (1995) and information pamphlets/packs published by curriculum authorities (contact QCA or SQA and see, e.g. QCA, 1998).

- Engaging the active support of people in the media. For instance, Prince William elected to study geography at A-level, in addition, Simon Jenkins (former editor of *The Times*), Michael Palin and David Putnam are strong proponents of geography. Look on the Internet for websites which may include information about famous people who have studied geography (see 'Websites' below).

- Reminding people, at every opportunity, that geography is the most popular non-mandatory subject at GCSE and is in the top six A-level subjects by total candidature in England and Wales.

Representing geography

My experience of campaigning for geography takes many forms. For instance, at a London High School's careers convention organised for 300 or so prospective A-level and degree level students and their parents, I was asked to represent geography. Each subject speaker was allotted ten minutes to make a case for their subject and/or career. There were ten presenters from professions such as engineering, accountancy and computer science. My presentation on geography was planned around four elements:

1. Questioning the stereotypes of the content of geography courses and the ways they are taught.
2. Giving a flavour of what modern geography courses are like and are about.
3. Describing some of the knowledge, understanding and skills gained through geography courses and related fieldwork.
4. Relating the stories of three female geography graduates, their subsequent careers and their

Pen portraits

Sue Cullum graduated in 1973 from University of Bristol with a geography and geology degree. She is now a consultant hydro-geologist for various companies with an expertise in quarrying and landfill. Sue advises on environmental matters especially water and gas pollution and occasionally lectures at Imperial College, London.

Landform analysis and interpretation and physical geography are important in Sue's work and her information technology skills have proved invaluable. Much of her degree work, therefore, has had direct applicability to Sue's profession. This is true of other environmentally related careers.

Gill Brown is now a reporter on an East Anglian local radio station having graduated in geography from the University of Hull in 1992. Her background in economic, political and social geography has proved invaluable, for instance, in the area of job losses and government regional grants. In addition, skills developed at university - communicating a case study, and collating and analysing data have proved helpful in Gill's professional life.

Jane Hargreaves graduated in geography from Cambridge in 1990 and became a member of the Institute of Chartered Accountants in 1993 having worked at Ernst & Young in the City. Jane now works as a corporate financier at the bankers Kleinwort Benson.

Jane feels very strongly that knowledge of urban and industrial geography allied to a logical approach, report writing, analysis of data and IT skills (especially spreadsheets and modelling) all gained at university have greatly helped in her present position.

acknowledgement of the benefits of studying geography (see 'Pen portraits'). These were intended to be atypical but nonetheless impressive geographer's stories!

My presentation had the desired effect and afterwards parents and students inundated me with enquiries about A-level and undergraduate level geography courses. This suggests that as well as training good quality teachers, our former students may well be our best advert!

Websites

The Association of American Geographers (www.aag.org) is planning a web page featuring details of people who studied geography. This will include, for example:

- Michael Jordan was a geography major at the University of North Carolina;
- Mother Theresa started her career as a geography teacher in Calcutta;
- a physical geographer, Justin Wilkinson, trains NASA astronauts to interpret landscapes from space;
- geographer Barry Bishop and his son were the first father and son team to successfully climb Mount Everest.

An equivalent list for the UK could be devised!

As far as employers are concerned you can quote Sam Toy, former managing director of Ford UK (and a geography graduate), who once remarked to Professor Peter Haggett (Department of Geography, University of Bristol) that geography:

'is the best subject in the world for getting on in industry'.

Promoting Geography in Schools

Chapter 2: Living geography

Garrett Nagle

This chapter illustrates the varied and dynamic nature of geography, and shows you how to raise students' awareness that geography is the route into some important issues. Case studies indicate how:

- local and national newspaper reports can be used to enthuse your students about geography in the classroom,
- journal articles and statistics can provide starting points for a topic which can be extended to other resources, such as CD-ROMs,
- linked classwork and fieldwork will help students develop an interest in their local environment, and
- predicting and anticipating events can be built into geographical studies.

The material in this chapter can be used at all stages of a geography course. Environmental issues, for instance, are equally relevant throughout the secondary phase, but you will need to adapt the content to the level you are teaching and your students' ability.

Using newspaper extracts

Geographical enquiries are often concerned with local issues - for instance, the impact of a new retail or housing development. Such developments are covered extensively by the local media and offer a good introduction to an enquiry. News reports will provide a variety of types of information - maps, photographs, statistics and graphics - all of which are familiar resources in the geography classroom.

The breadth of national and global topics covered in the media demonstrates the newsworthy nature of geography. The effect of El Niño weather systems, famine in North Korea, the British beef crisis, the closure of high street shops, storms and floods, and regional variations in wealth, unemployment, disease and mortality rates are some aspects of geography that feature regularly in the news. Some aspects of geography - the weather, for instance - are easily identified, whereas others are less obvious - for example, the geography of crime or agricultural issues. You should use these reports to demonstrate that if students keep their eyes and ears open they will realise that geography is happening around them every day.

The following activities will help your students get the most from the geography in the media:

- Monitor your local newspaper for two weeks looking for reports which have a geographical flavour.
- Classify the items you find as physical geography, human geography and 'issues' and file the extracts under these headings.
- Watch the TV news - both local and national - over a two week period and make notes under the same headings.

Then either:

- Choose one event, combine the TV and newspaper information you have gathered, and present it as a poster. This should explain the causes and effects of the event; a map would be very useful.

or

- Make a presentation to your class describing a geographical item in the news, from your area or nationally.

Younger students will need you to provide some background information to the topic; an example is given below. As their investigative skills progress, they can be encouraged to undertake their own research, from a variety of sources.

Example 1: Nitrate loads in rivers

The levels of nitrates in water have an impact upon the population consuming the water. The rise in the incidence of stomach cancer and blue baby syndrome and the threat to wildlife are causes for concern, as the following two extracts show.

THE NITRATE FACTOR
(Daily Mail, 9 September)

Scientists believe they have discovered a link between childhood diabetes and nitrate levels in tap water. The connection was found at levels well below the maximum permitted under EU standards for drinking water. Average nitrate levels ranged from 1.5-40mg/l (milligrams per litre) although the highest level was 59mg/l. The EU maximum level is 50mg/l. There was a significant link with the level of childhood diabetes in an area when the average level of nitrates rose above 14.85mg. The likelihood of developing childhood diabetes increased by 15%. About 20,000 children in Britain are diabetic and need regular doses of insulin to control blood sugar levels.

However, researchers warn that diabetes could be caused by another chemical and that high nitrate levels may coincide with its presence.

BP MUST PAY £7000 FOR PIPING SEWAGE INTO STREAM
(based on the Oxford Times, 6 September)

British Petroleum was ordered to pay £7000 in fines and costs after it admitted pumping sewage into a North Oxfordshire country stream, threatening wildlife. The company was allowing sewage containing up to six times the permitted level of ammonia into the waterway at Weston-on-the-Green. Human waste from toilets at the BP garage was insufficiently treated before it was allowed to pollute the small stream - a known spawning ground for trout. The sewage is no longer pumped into the stream and is being taken away by tanker instead.

Further information

Groundwater passes through soils and rocks to reach the water table, and different types of rock absorb the nitrates in the water at different rates. For example, chalk aquifers (water-bearing rocks) have a response rate (i.e. the length of time they take to absorb the nitrates) of 40-50 years, sandstone 15-20 years and limestone 4-5 years. Consequently over one-third of the UK population who get their water supply from chalk aquifers are likely to experience rising nitrate levels for the next 40 years or so. In Britain, legislation regulating the use of nitrate fertilisers on agricultural land has been introduced and 72 'Nitrate Vulnerable Zones', covering a total of 720,000 hectares, have been created. Despite these measures widespread

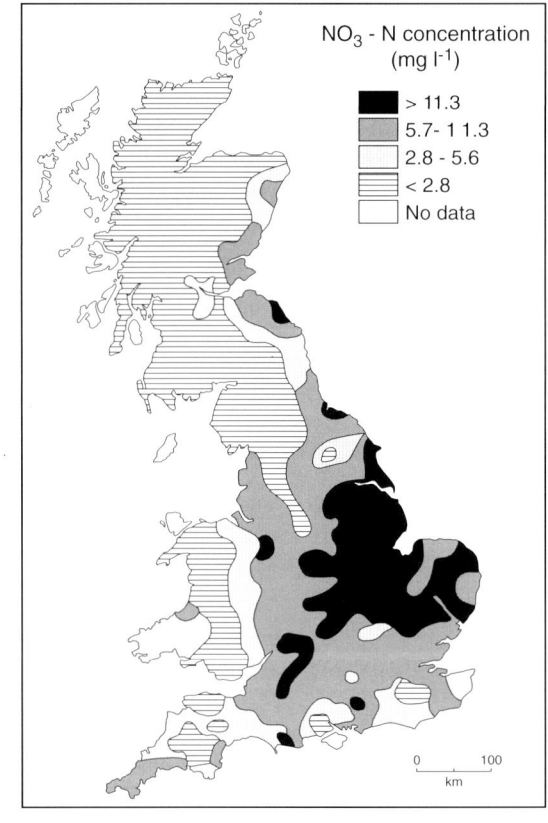

Figure 1: Mean annual nitrate concentrations in Britain. Source: Nagle, 1998.

concern remains regarding the rise of nitrate levels in rivers and groundwater. The mean annual nitrate (NO_3^-N) concentration in Britain is shown in Figure 1. Two factors are important here:

1. Upland areas in the north-west have high rainfall, diluting the nitrate to lower concentrations, whereas lowland areas in the south-east have lower rainfall and consequently higher concentrations of nitrate.
2. Lowland areas, such as East Anglia, which are centres of intensive arable farming, apply high levels of nitrate fertiliser per hectare, whereas upland areas are more likely to be used as pasture.

The long-term trend is for an increase in levels of NO_3^-N in rivers, with many exceeding the safe-limit of 11.3mg per litre set down by the EU. However, it is unlikely that drinking water will be treated, at least in the very near future. The main constraint to treatment remains cost: between £50 million and £300 million each year. Who should foot the bill - the government, the farmers, the manufacturers or the consumers?

Already over 5 million people live in areas with too much nitrate in the water. High levels of nitrates in water have been linked to diseases: stomach cancer and 'blue baby syndrome' (haemoglobinaemia - oxygen starvation in the gut of a young child). Scientists are not positive that the cause of these diseases is nitrates in the water. Some think it may be the increased level of phosphates (derived from agricultural practices); others think insufficient cases have been proven to bear out the statistics. Meanwhile the level of nitrates continues to rise.

Student activities
GCSE students can be asked to organise an enquiry into the question 'How safe is your water?'. This could start from key questions such as:
- Where does your water come from?
- How is it tested for quality?
- How is it monitored?
- How is it used?

You will need to contact the local Environment Agency office and water board and the council environmental health department. The school chemistry department might let students test samples of tap water and/or river water (for pH, nitrate, nitrites, aluminium, phosphates); you could also contact the manager of the local sewage treatment plant.

Using extracts from journals and statistical information

Extracts from magazines and journals/periodicals can be a good starting point for a study of, for example, the geography of health (see Figure 2). Ideally, the extract should contain local, regional and national data, and may concentrate upon one aspect of health, for example, health care provision in the local area, the spread of diseases within a region or national differences by gender in the number of deaths. This can be supplemented with more extensive (international) information, such as text, figures and graphs similar to that shown below.

- Proportionately fewer babies die in East Anglia than the rest of the UK: 5.2 deaths per thousand in the first year of life, compared with a high of 7.7 per thousand live births in Yorkshire and Humberside.
- Cancer registrations are highest in Scotland.
- Human immuno-deficiency virus (HIV) is present among the major risk groups throughout Britain, but highest in the Thames region.
- The highest death rates from heart disease are in Scotland and Northern Ireland.

Figure 2: Health in the UK. Source: Regional Trends, 1996.

Example 2: The geography of health and health care

Geographical investigations into health and disease have traditionally focused upon two main aspects: *disease ecology* or *epidemiology* and the provision of *health care*. Disease ecology is the study of why certain diseases are found in different geographic environments, for example, why there has been a high incidence of leukaemia around Greenham Common in Berkshire and around Sellafield in Cumbria. Studies of health care provision consider geographic, social and racial inequalities in access to health care. Most recently, *humanistic* and *structuralistic* approaches have influenced medical geography. The humanistic approach stresses human values and experiences, and attitude towards illness (a feature increasingly important in personal investigations). The structuralist approach stresses the political and social conditions which give rise to unequal incidence of illness and access to treatment. Essentially, the geography of health suggests that where you live affects your health and also the care you receive.

Nationally, the health profile of the UK has shifted over time from a prevalence of infectious or contagious diseases (*epidemics*) to a prevalence of those that cause a gradual worsening of the health of an individual (*degenerative diseases*). This is known as the epidemiological transition model. Economically less developed countries would be expected to have a high number of deaths and illnesses from diseases such as respiratory infections, measles and gastro-enteritis (diarrhoea and vomiting); an economically more developed country (EMDC) would be expected to have more deaths and illnesses from heart attacks, strokes and cancers, diseases which are not infectious or communicable. (The exception to this is the rise in auto-immuno deficiency syndrome (AIDS), and with it tuberculosis, in EMDCs in the last decade.)

Figure 3 shows changes in the causes of death in the UK. Deaths from influenza and tuberculosis, common in the mid-nineteenth century, are now quite rare, while the rise in cancers, strokes and heart disease is equally striking.

Mortality rates fell rapidly between the mid-eighteenth century and the Second World War, with those for women falling faster than those for men, but the basic pattern of mortality in England and Wales has since remained unchanged. The national average for mortality rates at the last census (1991) was 1067 deaths per 100,000 people, with only 3% of the population living in areas where the mortality rate was 1333 per 100,000 (or 25% more than the national average). The mortality rate was close to this figure in many northern cities, in Wales and in London. Mortality rates in more affluent areas, such as the Home Counties, at 853 deaths per 100,000 people, were 20% less than the national average.

There are a number of reasons why mortality patterns have changed over time:
- a decrease in infant mortality in the last 150 years
- during the 1980s suicide increased more than any other cause of death
- the last decade has seen an increase in the incidence of lung cancer and respiratory problems in the UK
- heart disease is on the decrease.

There are, however, regional variations between the causes of death.

Average 1848-72	
Infectious diseases	321
Tuberculosis	146
Scarlet fever	57
Typhoid	38
Respiratory diseases	148
Bronchitis	66
Pneumonia	57
Diseases of the:	
nervous system	129
digestive system	83
circulatory system	53
Other causes	266

1991	
Diseases of the circulatory system	472
Heart diseases	264
Strokes	177
Cancer	270
Respiratory diseases	155
Bronchitis	48
Other causes	137

Figure 3: Principal causes of death in the UK, rates per thousand population, 1848-72 and 1991. Sources: Beaujeu-Garnier, 1978; Regional Trends, 1996.

Student activities

The publication *GeoActive 163: The Geography of Disease* (summer 1997, Stanley Thornes) provides plenty of data for studying disease at a regional level. 14-16 and 16-18 students can also use CD-ROM information, for example, *The 1991 Census CD-ROM* (HMSO), to undertake studies similar to those outlined below.

For your chosen area (urban or rural) collect data from its wards (small-scale electoral areas).
- Find out the percentage of people in the local area who suffer from long-term illnesses. Divide the population of each ward by the number of people with long-term illness to find out those wards with the highest long-term illness (percentage of total).
- Identify the wards with the highest and lowest rates of long-term illness. To what extent are these wards perceived to be 'deprived' and 'affluent'?
- Look for indicators on how the type and number of long-term illness varies with age.
- Use the information shown in Figure 3 to describe the changes in the disease pattern in the UK between 1848-72 and 1991. Graph some of the diseases to make direct comparisons (e.g. Figure 4). Attempt to answer the question 'Why do disease patterns vary with levels of economic development?'.
- Using data from the *1991 Census CD-ROM* (on factors such as unemployment, ethnicity and/or housing tenure) decide whether there is any correlation/relationship between long-term illness and socio-economic conditions.
- Conduct an environmental quality index to see whether areas of high long-term illness differ from those of low. One such index is shown in Figure 5 (but you could add a range of other categories). The categories score 1-5 (1 being the lowest, 5 being highest) and sum these figures to reach an index for that area.

Category	Low (good)	High (bad)
High density housing		4
High use flats	1	
Graffiti	2	
Signs of vandalism	2	
Wasteland/ derelict land	1	
Noise		4
Total	14	

Figure 5: Sample environmental quality index.

Linking classwork and fieldwork

Linked class and field activities will help students learn about the factors that affect people in their locality. You could undertake a study of local crime, for example; most people will at some time, either directly or indirectly, be the victim of crime, and all of us have strong views about criminal activity. These views may be influenced by television shows, such as *The Bill* or *Inspector Morse*. To find out more about crime in your area invite the Police County Liaison Officer and/or a representative of the Neighbourhood Watch Scheme to discuss the matter with your students. Your students can then undertake field interviews with members of the public (or security guards), plot the location of closed circuit televisions and use statistical analysis of police data to prepare reports on crime in the local area. Students can use data relating to car theft, for example; it is more reliable than for other crimes because car insurance claims require thefts to be reported to the police. If they are not, the victim cannot make an insurance claim.

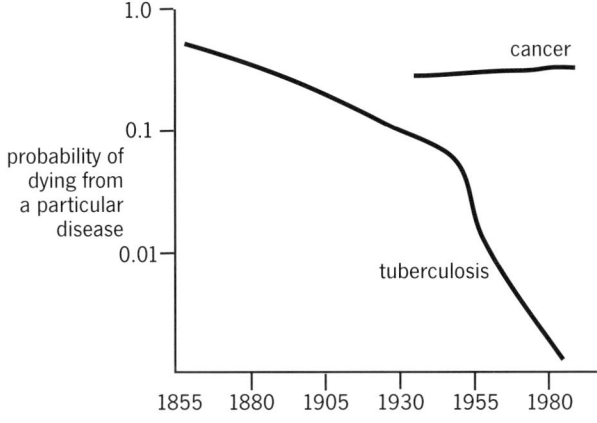

Figure 4: Changing patterns of disease in Britain: Tuberculosis and cancer. Source: Review of Registrar General on Deaths in England and Wales, 1941-90.

Example 3:
The geography of crime in Oxford

In and around Oxford, vehicle crime is widespread (Figure 6), including:
1. theft of cars, vans, mopeds and motorcycles, etc.,
2. theft of parts from the vehicle, and
3. attempted theft in either of the above categories.

In Oxford in 1994 over 6000 cases of vehicle theft were recorded by the Thames Valley Police; more than 25% of all recorded crimes. The pattern is concentrated in the Headington area and to a lesser extent around the central area, parts of East Oxford, Blackbird Leys, the University and North Oxford near the Pear Tree park-and-ride car park. A number of factors can help explain this pattern:

Long-term car parks are the focus for much car crime in Oxford. Photo: Garrett Nagle.

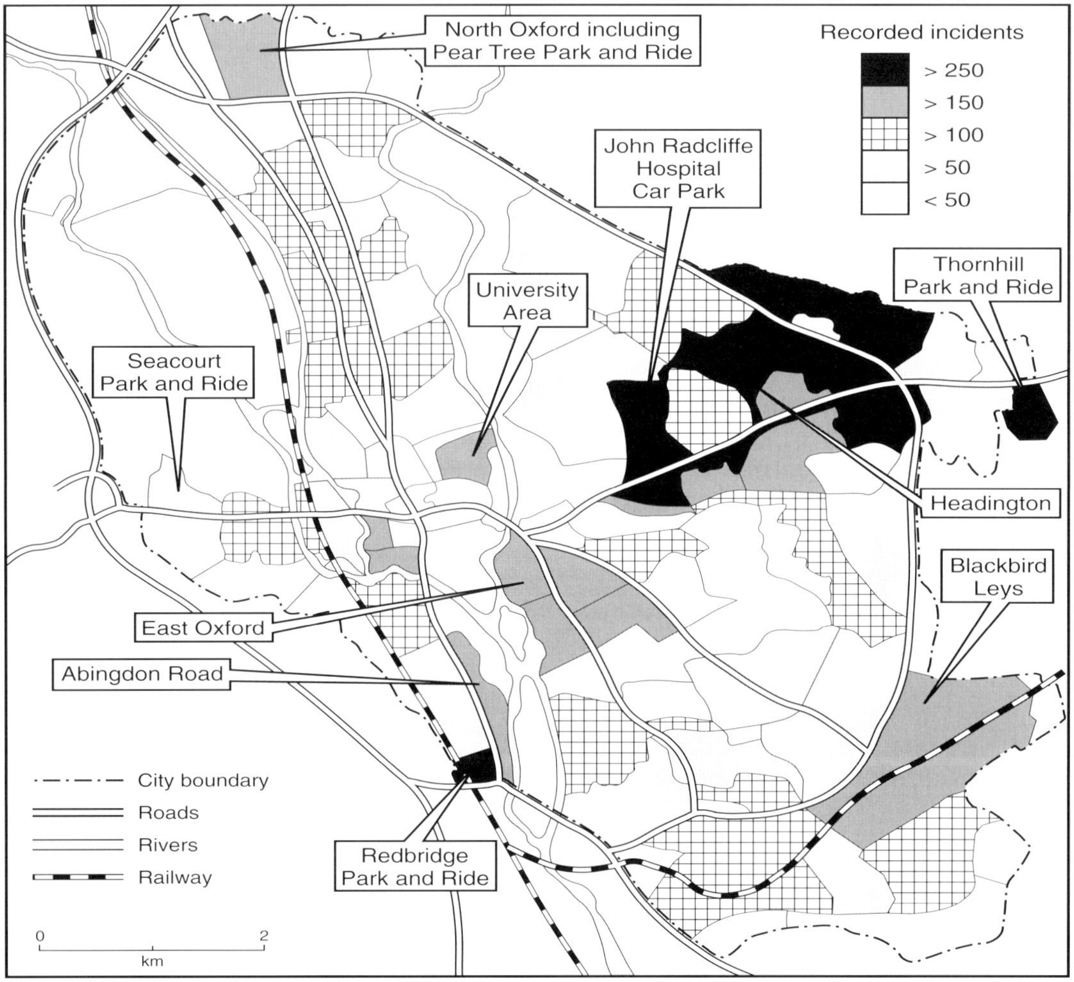

Figure 6: Vehicle theft in Oxford. Source: Nagle, 1997.

Promoting Geography in Schools

Figure 7: Annotated photographs produced as part of a project on crime in Oxford by Mia Brunner, St Edward's School, Oxford.

- The location of car parks (see photograph), especially long-term ones such as park-and-ride schemes at hospitals and places of employment.
- Areas where joyriding is not uncommon (Blackbird Leys and to a lesser extent Lower Wolvercote).

Increasingly, extra security in car parks is being used to reduce levels of car theft.

Student activities

As well as the activities outlined above, 11-14 year old students can visit local park-and-ride schemes to gather evidence by using questionnaires. They can ask the users such questions as:
- Do you think the park-and-ride car parks are safe?
- Would you be prepared to pay more to park here if this meant better security?

14-16 year old students can carry out a photographic transect of an urban or rural area. At each stopping point they should photograph one or two features, make notes on the area and record signs of crime or its impact. Back in the classroom, students annotate the photographs on a sheet of A4 paper by labelling around the images, using their field notes as reference (see Figure 7) (see also Speake and Donert, 1998).

Using data from the local police (for instance, Thames Valley Police offer statistics on the incidence of certain crimes) and information from neighbourhood watch schemes and security firms, 16-19 year old students can create a report on the local area. They should undertake a field survey on the localities of closed-circuit television cameras used by councils, the police and security firms. Their project work should include photographs, for instance.

Looking forward

'With careful selection, and linking of places and themes, it is possible to build "predictable topicality" into [the] long-term curriculum' (Stanfield, 1995, p. 169). Stanfield (1995) also provides examples of themes which can be integrated into your existing programme of study:
- the nature, causes and effects of tectonic processes integrated into a study of Japan,
- the causes and effects of river flooding integrated into a study of a region of the UK,
- aspects of weather and climate integrated into a study of Central America.

The latter is very topical: environmental disasters are widely reported in the media. Geographers often look forward to reports on environmental disasters with a macabre sense of duty - 'a new case study'. A case study of one environmental disaster can offer the impetus for a range of classwork, using a variety of resources (including information and communications technology).

Example 4: Hurricane Mitch

In November 1998, Hurricane Mitch devastated Central America. Parts of Honduras experienced over 1000mm of rainfall in five days. The deluge filled the cone of the dormant Casita volcano in north-western Nicaragua, causing it to burst through the side of the mountain. A wave of water, mud, rocks and trees crashed down on four villages, burying everything in its path and killing an estimated 15,000 people. The tragedy of Casita was the most spectacular single disaster inflicted by Hurricane Mitch, but it was not the only one. Most of the damage came from day-upon-day of heavy, continuous rain which caused huge floods and many mudslides across Central America. In Honduras and Nicaragua bridges, roads, power lines, plantations, crops and cattle were swept away. Hundreds of thousands of people lost their homes. About one-third of Hondurans were affected by the flooding. With half the population aged under 25 and the infrastructure, including many schools, destroyed, the country is facing the possibility that an entire generation of Hondurans may grow up without an education.

Up to 70 per cent of Honduran economic output was lost. The 1998 and 1999 economic cost of the disaster may total US$1.5 billion (£903 million). Rebuilding the country's infrastructure, i.e. transport, communications links and services, could cost US$2 billion.

Farming is vital to both Honduran and Nicaraguan economies, and almost half of the losses are in the agricultural sector. 25% of Honduran coffee plantations were destroyed; 75 per cent of banana plantations; all shrimp farms have been buried under mudslides. Sugar and citrus fruit plantations have been hard hit. The Sula Valley in northern Honduras, whose factories and farms account for some 60 per cent of the country's GDP, was completely flooded. Managers of Nicaraguan coffee plantations put the destruction at 20%; newer crops such as oranges have been wiped out. In both countries, many small farmers have lost everything. Coffee growers say the 1998-99 crop fell some 20 per cent short of expected yields. As Central America produces 10-12 per cent of the world's coffee this has implications for the price of coffee on the world market.

Student activities

Students of any age could undertake the following activities based on studies of natural disasters:
- Collect information from a variety of sources, e.g. newspapers, weather forecasts, television news, Internet, geography textbooks, to compile a disaster scrap book.
- Create a 'disaster map' which shows examples of disasters in different locations. Photographs are often available on CD-ROM/the Internet: print those which show the disaster or its impact and relate them to the map. Display the finished poster(s) in the classroom.
- Set up a newsroom. Use software packages such as *Newsroom Extra* and 'role cards' for each student: runners, researchers, graphic designers, 'photographers' (who select images from the Internet and newspapers/magazines), writers, proof readers. Material can be prepared, stored and printed out as newsflashes (and/or 'reported live'). Predetermined intervals between each newsflash can be used to regulate the pace of classroom activity.

For 16-19 students you could:
- Arrange for a speaker from your nearest university geography department, or one of the aid agencies, to talk about natural disasters. Students should devise questions for the speaker. Small groups of students could follow up by producing a report on the work of the university or aid agency.
- Ask students to find out about long-term consequences of the disaster - media interest often peters out after the first few days. One aspect of this is the effect of one disaster on the global economy, through a shortfall in the supply of coffee, for instance.

Other ideas

The scope of geography is limitless, which is a great strength, but can also be a problem: it can be difficult to keep up with everything all the time. Jeff Stanfield's article 'Topicality in geography teaching' (April 1996, *Teaching Geography*) should give you some pointers as to how to keep your classroom geography fresh. Excellent sources of ideas, classroom material or case studies, which can be used directly or adapted, are:
- *Teaching Geography*
- *Geographical Magazine*
- *Geofile*
- *GeoActive*
- *GeoPress*

The Internet is also extremely valuable; encourage your students to take advantage of it. One of the best sites is the University of Leicester - http://www.geog.le.ac.uk/cti - which is organised by topics. This offers one of the clearest routes to up-to-the-minute geography. Try it and see!

Conclusions

Ordinary and extraordinary events, from local to global, can have geographical significance, and geography can explain where and why such events take place. This chapter also demonstrates how students' natural curiosity and own experiences can be channelled and developed in the geography classroom. A variety of resources, together with teacher or student-resourced information, can provide relevant, topical and interesting geographical studies, and researching factual data on a topical event is a very positive learning experience. Using these approaches will encourage your students to continue their study of geography beyond your classroom.

Promoting Geography in Schools

Chapter 3: Tailoring your approach to your audience

Tim Price-Walker with Peter Fry, Keith Grimwade and Alan Marvell

In this chapter Tim Price-Walker describes how a typical geography department in a secondary school can 'sell' geography to different 'markets': current students, prospective students and their parents, together with contributions from Peter Fry, Keith Grimwade and Alan Marvell. They also make recommendations about maximising the use of resources to promote geography as a worthwhile subject in the curriculum.

The educational world is very dynamic: we cannot afford to be complacent. Making geography a marketable and enjoyable subject, both now and in the future, depends very heavily on the ideas and initiative of geography teachers.

How one school marketed geography

Langley Park School is a large single sex comprehensive in Kent with over 1350 students. It was recently awarded Technology School status, reflecting its vision of information and communications technology being at the heart of future education. One result of this change in status was a cut in teaching time for history and geography from 1 hour 45 minutes to 1 hour 10 minutes per week. There have also changes in exam syllabuses and staffing of the department. This new situation called for a re-appraisal of how we perceive geography as a subject, and therefore how we should market it to students into the new millennium. By taking the initiative in developing new approaches to geography, the geography department has managed to avoid the potentially negative impact of these changes.

Creating a 'marketing plan'

Central to our vision of geography is that students' experience of the subject should be enjoyable. Instead of viewing the loss of teaching time as a problem, teachers at Langley Park saw it as an opportunity to concentrate the available time on the topics young people want to learn. Equally, we wanted to ensure that we marketed the subject effectively, and that our marketing of geography reflected the flexible, dynamic nature of modern education.

If you are selling something - whether cars, tins of baked beans or geography - it is essential to have a marketing plan. 'Keep things simple' was our watchword; at Langley Park School we confined our marketing considerations to four main points:

1. In terms of marketing the subject, where is the Geography Department now?
2. Where do we want to be in the future?
3. How do we intend to get there?
4. What are the relevant timescales for achieving this?

Using these questions we were able to draw up a comprehensive 'Plan of action' for the promotion and marketing of geography within Langley Park School. The teachers' plan was centred on three main points:

- Identifying potential markets, and tailoring our marketing to them. For each audience, we considered a range of questions, for example, 'What are students interested in?, 'What do parents need to know?'.

- Identifying initiatives which had worked well in the past and building on their success.

- Exploiting the rich variety of resources to market the subject.

Promoting Geography in Schools

Selling to the right audience

Our potential market splits into three groups: current students, their parents, and prospective students and their parents. Each group requires a different strategy, to be employed at different times.

Current students

You are the best advert for your subject; promoting geography to this group means making yourself available to them as often as possible, providing interesting classroom experiences (Chapter 2), perhaps starting a geography club (Chapter 6) and encouraging them to look beyond their studies to possible employment (Chapter 5). Your obvious enthusiasm for geography may even encourage students to contemplate teaching the subject.

Current students may range in age from 11 to 18 years; the best marketing will target each year group, but in the frenetic world of teaching this is unlikely to be feasible and you will have to prioritise.

Geography related vocational qualifications

If your department offers GNVQ Leisure & Tourism as one of a wider range of vocational courses at 16+ promotional material can be used to reflect the vocational aspects of geography. It can focus on the skills and potential for employment in various industries. Where GNVQ/Group Awards and

- Experience clearly points to the fact that to ensure students like and continue to like geography, they need to enjoy it 'from the start'. This makes year 7 vitally important. Keep them interested and stimulated - field trips are difficult with younger students, but if a day trip can be arranged, just observe the results! (Often it will be talked about right through to age 16!) It is so much easier to teach less obviously interesting topics in geography later on in their school career if they know that geography is and can be enjoyable!

- You may find it beneficial to ensure that the geography topics and the quality of the lessons at age 13-14 are particularly worthwhile and enjoyable, shortly before they decide on their subject options for GCSE/S-grade (e.g. Natural Hazards and Environmental Issues). Altering the year 9 scheme of work can encourage student to take geography as an option, giving them a positive feel for the subject.

- Focus on two important watersheds: at 14, when students are just about to choose their options, and at 16, when they are considering further education or prospective careers.

- Interview every 14-year-old student (briefly!) about their options. Students respond well to one-to-one communication, and it also provides advance information about GCSE/S-grade take-up of geography. Do the same with 16-year-old students interested in A-level/H-grades in geography. It is time-consuming, but very good public relations.

- Teachers often abandon 16-year-old leavers as a 'lost cause', but this is short-sighted: flexibility and re-training are now important factors of most careers, and positive school experiences of geography make it more likely that students will return to it in later life.

- Plan displays so they promote geography to particular age groups at the most appropriate times. For example, a year 9 options board can be displayed just after Christmas, when students first consider their options. You will need to put up displays for 16-year-old leavers, and prospective A-level/H-grade students, towards the end of the year.

- At 16-18, geography competes for students with GNVQ/Group Awards subjects and currently fashionable A-level/H-grades such as Media Studies, Business Studies and Psychology. Promotion at this level can be partly geared towards the practical applications and direct relevance of geography to the student.

- Remember to talk to older students, even if you are not teaching them. These 16-18 year olds could be future geographers! Encourage those who wish to pursue geography-based studies. Telling smaller groups of A-level/H-grade students about your own undergraduate experiences of geography can be a worthwhile activity. Alternatively, you could ask a former student to do so.

geography is to displayed side-by-side at open evenings, the 'fieldwork' aspects of GNVQ/Group Awards should feature prominently, particularly the research, analysis and report writing that is required.

To promote GNVQ/Group Awards courses successfully the need for targeted marketing is crucial. Emphasise that students take responsibility for their learning and progress and it should attract a wide range of interest. Promotional material (colour leaflets and posters) for GNVQ/Group Awards is available from the awarding bodies RSA, BTEC (now part of Edexcel), SQA and C&G and also from QCA/SQA. Your departmental leaflets should clarify and simplify the jargon associated with these qualifications. Any such leaflets should emphasise:
- the titles and descriptions of units studied,
- visits, guest speakers and local industry links,
- grading and assessment, and
- employment possibilities and progression routes.

Alternative qualifications to carry a vocational theme include GCSE Travel & Tourism (Marvell and Smyth, 1996) which shares some commonality with the GNVQ Intermediate Leisure & Tourism Programme.

In schools and colleges where A-level/H-grade Geography is seen to be competing with GNVQ/Group Awards, similar advice is given. Many teachers consider that the market for GNVQ/Grop Awards students is different from that of the 'traditional' A-level/H-grade qualification.

Parents of current students

Your best opportunity to talk in detail to parents of current students is at parents' evenings.

Parents' evenings

These can open and maintain a dialogue with parents at all levels of their children's geographical studies. Parents attend these events to learn how their children are doing in the current year, but you should not neglect the opportunity they offer to promote the subject. You could consider engaging parents with their children's subject by recruiting their help on geography field trips and by keeping in contact with parent-governors who influence school management decisions.

Make sure you have a departmental leaflet available. This can be one sheet of A4 folded in half,

Talk to parents at open evenings. Photo: Tim Price-Walker.

specifically aimed at parents. It should briefly explain the reasons for studying geography and the career opportunities it offers (see Figure 1 overleaf, for example, and the GA's new *Careers Pack* (Palôt, 1999)). This can help parents to encourage their children to choose or continue to study geography.

Parents' evenings usually include discussion of students' GCSE/S-grade options. The school's 'Options booklet' will include a page or more on geography, which must fulfil two inter-related purposes: to remind students of the reasons for taking geography, and to explain to parents why geography is a good subject to study at GCSE/S-grade. Both of these aims require straightforward language, clear presentation and a 'welcoming' style. A possible model for constructing a 'user-friendly' entry for geography is shown in Figure 2.

Other opportunities to promote geography to parents of current students

If you are asked to undertake a presentation at a meeting which includes parents ensure that you include some 'audience participation' to keep attention and increase interest. You could try one of the following:

- A GCSE/S-grade examination question. Invite responses and mark the parents! This is a marvellous activity for showing parents the topicality of geography and helps them to empathise with the sort of work their children undertake in the classroom.

You can help your child by:

- providing a working area at home where your child can work on large and small geography projects relatively undisturbed;
- being prepared to offer more help with some geography projects such as fieldwork trips;
- keeping in contact with school at open days, etc., so that you know what is going on;
- reminding your child to meet the deadlines set for geography homework.

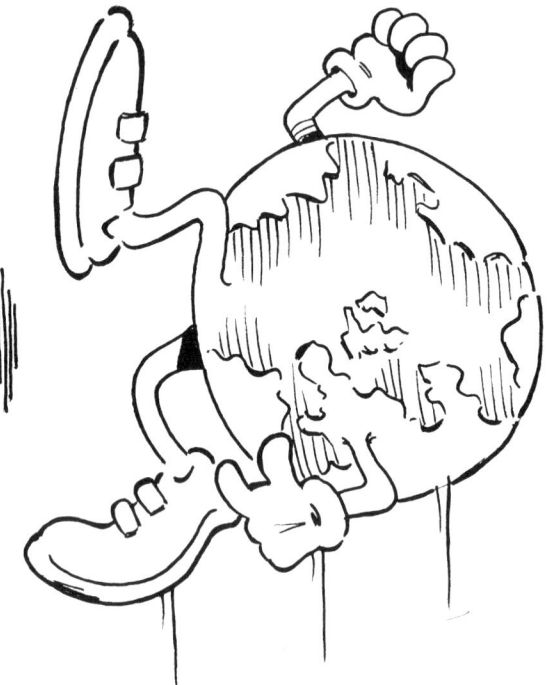

Geoging keeps the world in shape!

THE GEOGRAPHICAL ASSOCIATION

Should my child do GCSE/S-grade Geography?

Where can geography take my child?

You should support and encourage your child to choose geography if he or she already enjoys the subject. If that's the case, read on …

14-16 geography in this school

What's the point of geography?

Geography is part of everyday life: world weather and food costs; multi-national companies and politics; population movements and urban and rural change; maps and holidays. Understanding these things is what geography is about. The levels of interest span all scales, from a local planning enquiry to global warming.

What will my child gain from geography?

Geography is the study of where places are, what they are like, what life is like in them, and how and why they are changing. It can help young people to:

- Read and use maps, atlases and diagrams.
- Analyse and evaluate data.
- Develop decision making skills.
- Gain a knowledge of the world and understand current events.
- Appreciate different cultures, in this country and abroad.
- Become aware of physical and human environments.

Studying geography widens children's horizons and helps them become better world citizens.

Geography studies the parts other subjects cannot reach!

Where will geography take my child?

Geography contributes to employability. Employers like geography qualifications because they demonstrate a wide range of skills, including data collection and analysis, computer literacy, word processing, self-motivation and teamwork skills. Some jobs for which GCSE/S-grade Geography is a requirement include:

- surveying
- town and country planning
- civil engineering
- landscape architecture
- transport and tourism
- estate management
- cartography
- exploration

For other careers, geography is an advantage:

- pilot
- lorry driver
- estate agent
- police officer
- courier

Coursework and field studies

Examination

Parents

Figure 1: An example handout for parents.

What is GCSE Geography?
In this opening paragraph information can be drawn from the GCSE Criteria for Geography (SCAA, 1995), to which all syllabuses conform. Focus on a small number of key features, using bullet points, each with an example:

> 'GCSE Geography is about topics such as rivers, cities and raw materials; it is about places - the local area, the UK, the European Union and other countries from around the world; and it is about issues such as pollution and global warming. It helps us to understand the world about us; how people in different parts of the world depend on each other: and how we all rely on a healthy environment.'

Why GCSE Geography is a useful subject to take
In this paragraph some of the above points should be reinforced and new ones can be added.

> 'As well as being an interesting and relevant subject in its own right: the skills learnt as part of GCSE Geography, such as map reading, are useful 'life skills'; employers are aware that GCSE Geographers know useful information about today's world, can find things out for themselves, can use maps, graphs and statistics and have learnt important skills such as report writing; and GCSE Geographers can go on to study a wide range of post-16 courses.'

What is special about the syllabus the school follows?
Insert the syllabus information here but not in great detail. Most students will base their decision on your school's key stage 3 course; their year 9 experience; what their peers say about the subject; and the extent to which the points made in the previous paragraphs strike a chord. However, they are often concerned about the type, length and style of coursework as they attempt to 'balance' the demands of their subjects.

How you can find out more
Keep the lines of communication open.

> 'If you, or your parents, would like to find out more about GCSE Geography, your geography teacher, or any member of the geography department, will always be happy to talk to you.'

Include a student's quote
One year 11 student's view of GCSE geography:

> 'I found geography really interesting because it explains why the world is as it is - how landscapes form, how cities grow and the causes and effects of environmental issues. It also made me realise not only that the world I have experience of is just a small part of the whole world but that our lifestyles - for example our extravagant use of energy - can have global consequences.'

Include a judicially chosen cartoon, illustration or icon. This can help a piece of text and real geographers usually include some sort of illustration on every page! As a final stimulus and focus of debate for parents and students discuss a poem such as the one shown below (this could be included on a separate sheet of paper).

Geography lesson
> When the jet sprang into the sky,
> it was clear why the city
> had developed the way it had,
> seeing it scaled six inches to the mile.
> There seemed an inevitability
> about what on ground looked haphazard,
> unplanned and without style
> when the jet sprang into the sky.
>
> When the jet reached ten thousand feet,
> it was clear why the country
> had cities where rivers ran
> and why the valleys were populated.
> The logic of geography -
> that land and water attracted man -
> was clearly delineated
> when the jet reached ten thousand feet.
>
> When the jet rose six miles high,
> it was clear that the earth was round
> and that it had more sea than land.
> But it was difficult to understand
> that men on the earth found
> causes to hate each other, to build
> walls across cities and to kill.
> From that height, it was not clear why.
>
> Zulfiker Ghose

Figure 2: A model for the geography page of your school's options booklet.

- Teach a brief geography lesson to parents. This provides an ideal opportunity for their participation in part of a 'typical lesson' and can involve parents offering feedback. It should focus on the changes in the subject of which parents are generally unaware.

- Look at a local issue. One general meeting on the issue of Heathrow's Terminal 5, almost became a mini public enquiry! Langley Park sixth-formers made a short presentation on the views surrounding this issue and parents were asked to clarify these views and why they were held - values analysis and great fun! Local issues are certainly not in short supply.

Such presentations are often more effective than speeches and leaflets. They give parents an idea of the geographical content, understanding and skills their children are developing and how geography relates to other subjects and the world of work. Finally, it offers parents the opportunity to respond to the subject and to how their children are taught.

Parents at home

Don't neglect parents who are unable to attend open days/evenings or general meetings. The kind of things you can send home include:

- leaflets/booklet for pre-intake open evening (Figures 1 and 2)
- leaflets detailing 14+ and 16+ options
- regular department newsletter
- details about the Geography Club (see Chapter 6) and events outside school (Chapter 7).

Prospective students and their parents

Both prospective students and their parents will attend open days, and whilst these events are intended to promote the school, rather than the subject, the geography department can play an important role in both promoting a positive image of the school and influencing future options choices. A student-produced map can be handed out to visitors as they enter the school.

Open days

The key word for successfully promoting geography and geography-related GNVQ/Group Awards at these events is **interaction**. Geography is such a visual subject that wall displays are an obvious choice, but remember that you will be in competition with subjects which are very colourful and active (PE, Drama and Music) or which have equally interesting resources (Technology, IT and Science).

The way to attract parents to geography is to stress its career-related benefits, and to tell them about good examination results and routes to A-level/H-grade and GNVQ/Group Awards; students will be drawn to demonstrations by current students and opportunities to use information and communications technology and CD-ROM packages, handle artefacts, etc. Encourage parents to use the CD-ROM packages too: they are unlikely to have had this experience when they were at school and it may enhance their understanding and image of the subject.

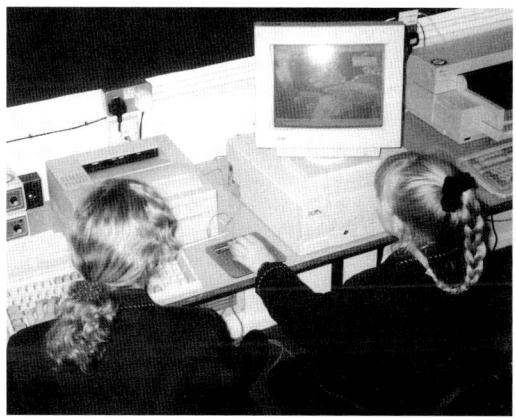

Students can help demonstrate ICT in geography to prospective students and their parents at open days. Photo: Tim Price-Walker.

You will get parents and prospective students to enjoy geography through an 'experience'. Often the simplest ideas have the most impact - we have found the following very effective:

1. Students and parents have come to see you, so it is important to be friendly and relaxed. Welcome prospective students and parents but don't go overboard! Allowing them the freedom to explore, with occasional encouragement to try out some of the activities on display, will usually

Promoting Geography in Schools

break the ice. Have your schemes of work on hand ready for any awkward questions which may arise!

2. Try to arrange your space as an 'exhibition area'. Designate room for displays, and ensure your visitors have plenty of space to move freely around displays or working models. Move desks together to create large display areas and place surplus chairs underneath out of the way: this doesn't take long and looks more professional.

3. Get students to hand out information flyers at the school entrance (ask the Head's permission first) to encourage parents and prospective students to visit the geography department!

4. Student displays should be thematic and colourful - and recent! Signs can be quickly word-processed and laminated for a professional look.

5. Computers should be switched on, with relevant software running, and both students and parents encouraged to explore the programs. The importance of giving parents a positive impression of the geography in your school should not be underestimated: they can influence students' decisions about the subject later on!

6. Aerial photographs of the local area are often a good focal point. Parents usually try to locate their homes. Two volunteer students could politely ask for the name of the road in which parents live and using a base map create a 'sphere of influence' map of parents on the evening.

7. Lay out examination papers with a sign saying 'So you thought you were good at geography? ... then try some of these!'. It is important that parents are made aware of the standards expected of students. Conveying this in a humorous way is less likely to frighten them!

8. Lay out student coursework material. Often this is one of the best ways to demonstrate student's individual work. Parents are usually impressed with the quality of GCSE/S-grade and A-level/H-grade projects, not least those based on the local area.

A 'cover conscious' display of student books can have many benefits. Photo: Tim Price-Walker.

9. At Langley Park School a competition is held at the beginning of the year (for each year group) to find the best covered exercise book based on a geographical theme. A substantial prize is offered! The results have been stunning. This also provides a wonderful array of geography exercise books for display on open evenings! Students respond to this competition indirectly by taking care of their geography books.

10. Ensure that exercise books are marked up-to-date with appropriate comments and targets. Parents respond very well to being allowed to freely view student exercise books. If you have

the confidence to allow visitors to peruse exercise books it will boost their confidence in your teaching!

11. A television and video recorder or an audio cassette can transform the environment of your geography room! Playing a video of volcanic eruptions or the rainforests of Brazil can contribute to the atmosphere. Programmes which include moving images and ambient

Pro-active displays provide a variety of stimuli on open evenings. Photo: Tim Price-Walker.

sound work well, attracting much more attention than those with lots of commentary. Background noise is also beneficial in avoiding awkward silences, especially when there are just a few people in the room! It will encourage your visitors to relax more. Get 16-18 students to record a video of a recent field trip, demonstrating that geography can be fun!

12. Use a presentation package (e.g. *PowerPoint*) to combine photographs, video clips, music and text as a selection of slides. These can illustrate a particular aspect of geography, e.g. coursework schemes of work. The resulting images can be automatically projected onto a screen or through a television. This kind of medium easily portrays geography as an exciting subject. (For further details, see Rogers, 1998.)

13. Display any working models the department has, or computer programs which can run automatically. Such ideas could include:

    - River channel simulation
    - Satellite imagery animation of the days weather
    - The present weather from an automatic data logger

14. Fill the black/white board, often a large vacant space on open days, with the scheme of work for each year. Briefly list the topics of study and fieldwork visits.

15. Word-processed and laminated direction signs to the geography department is a highly practical way of promoting the usefulness of geography!

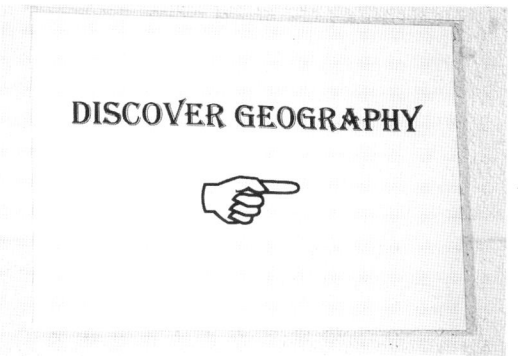

Photo: Tim Price-Walker.

16. Put out copies of the curriculum resources including textbooks currently used for each of the year groups.

17. Use volunteer students to demonstrate computer programs or to discuss their geography course with prospective students and their parents. Students can act as guides to parents who wish to be escorted round the department.

Figure 3, overleaf, provides a geography checklist to ensure that you/your head of department includes all of the above.

Promoting Geography in Schools

Open evening checklist
Room layout ☐

Static display
- Textbooks ☐
- Students' exercise books ☐
- Students' projects/coursework/fieldwork ☐
- Student's models/wall display ☐
- Globes/atlases ☐
- Ordnance Survey maps ☐
- Posters ☐
- Slides, audio-visual material ☐
- Worksheets, resource sheets, support materials ☐
- List schemes of work on the board ☐
- Artefacts (geological materials, etc.) ☐
- Local area materials including: ☐
  - geological materials
  - historical maps
  - maps at various scales
  - photographs, press cuttings

Active
- Display examination papers, ask parents to answer one question ☐
- Working model, e.g. river simulation ☐
- Computer software package, interactive CD-ROM if possible ☐
- Magnifying glass with aerial photographs/maps of local area ☐

Ask students to:
- Demonstrate how they use ☐
  - fieldwork equipment ☐
  - software/CD-ROMs ☐
- Answer questions ☐
- Act as guides ☐

Other considerations
- Printed handouts (see Figure 1 and the GA's new careers pack) ☐
- Past GCSE papers (challenge parents and offer a prize for the best mark of the evening) ☐
- A quiz (utilise GA *Worldwise Quiz* and *World Puzzle* books) with prizes for the most correct answers ☐
- Careers leaflets from GA careers pack ☐

Figure 3: Open evening checklist for Head of Department.

Maximise your assets

As there is so much pressure on teachers' time it makes sense to be efficient (the checklists should help). There's no point constantly re-inventing the wheel: if they worked well in the past, recycle your department's 'old' ideas and resources. For example, historical maps are great for showing how much the world has changed in a short time. Parents and students often become immersed in comparing the historical map with a recent one (particularly ones of the local area): a practical opportunity to demonstrate that geography is a dynamic and interesting subject!

Resources

Your first, and most important, resource is your classroom, which can provide a solid foundation for all your geography promotion. Its appearance is vitally important, and can offer a strong positive image of the subject as well as a rich learning environment. Primary schools often provide stimulating classroom environments whereas secondary classrooms are often barren by comparison.

At Langley Park the staff wanted students to feel positive about entering the geography department, and we did this by providing the geography department with its own unique and pleasing identity. In the same way that territories are important to animals, so they are to humans; we worked on the premise that students would perform better in their 'own' territory. Geography form rooms were created; classes registered in the geography rooms encouraged student involvement in the decoration and maintenance of the rooms. This simple psychology greatly enhanced the promotion of the geography rooms: by encouraging students to decorate the room, with the minimum of supervision, we both involved them in 'ownership' of the rooms and gave the room a geographical identity. The outcome has been remarkable - one form group became extremely protective of their environment and was very upset when posters were ripped or went missing! Wall displays are periodically updated and revitalised, so students continue to look and learn. Some pointers for enhancing the geography environment are given below.

Promoting Geography in Schools

- A 'Geography in the news' display, near the entrance, is essential - up-to-date information ranging from field trip news, through examination advice, to the latest world tremor (see photograph). Often the geographical content does not have to be paramount: we want to foster the notion that the department is keen to keep students up-to-date with current events.

Keep students up-to-date, ensuring that geography is perceived as a particlarly interesting subject.
Photo: Tim Price-Walker.

- A balance of commercial, formal and students' own work in the classroom seems to work best, particularly when based on a theme in a particular year group.

- Laminating word processed signs and outstanding student work adds a professional look to wall displays

- Direction signs to the geography department always work particularly well, especially at the beginning of the year when year 7 students are always getting lost!

- Walls are not the only display areas: don't forget the ceiling and windows. Suspend mobiles or inflatable globes from the ceiling. Stick signs printed on overhead projector transparencies to your windows; they can indicate compass directions or simple geographical terminology.

- Cultural geography is making a comeback. Make geography a practical subject by ensuring your students realise how geography really does affect our lives. Media images from newspapers, magazines and advertising can reveal a surprising amount of real life geography (Chapter 2). Our school produced a multimedia display on unexplained natural phenomena linked with geography (see photograph). Exploiting the popularity of a current sci-fi series, *X Files in Geography* was a great success. An audio of the theme music helped set the scene.

Contemporary displays link geography with cultural trends. Photo: Tim Price-Walker.

- Use of resources is also very important. Keep a record or inventory of your geography resources. The resources we use to promote geography include:

  a) *Software packages:* This alternative means of study and research has been introduced into most geography departments, with CD-ROMs being most popular. Geography is at the forefront of CD-ROM development because it's visual nature makes it ideal for this medium. Animation, maps and text can be viewed on the screen or printed off at leisure.

  b) *Screen savers:* These can be accessed and a departmental message promoting the subject written (and protected by a password!) to move across the screen. (You could include the names of the geography department staff or lists of your geography CD-ROMs.)

c) *Networked computers and the Internet:* Some schools have networked computer rooms. In our school the geography department has ensured that geography and ICT are a compatible and unified medium. Within the network, geography has its own workstation area with geographical-based software suitable for all age groups. Students can access a 'domain' to research and word-process an assignment from anywhere in the school. The Internet is also becoming a widely accepted medium for learning. Make sure that geography students are aware of the advantages of using the Internet for geographical research! Create a list of the addresses of the top ten geographical sites to enable easy access. If the school has a web page, ensure that geography has a section and that the information provided is up-to-date and well maintained (a good task for a 16-18 year old perhaps!). You may well be surprised with responses you receive from geographers all over the world!

d) *The library:* This is an important work place within the school (see photograph). Do not ignore the marketing of geography in this area! A feature of our school library is a combined television and video with personal headphones. Students can book out geography videos to watch - a particularly useful resource for A-level/H-grade students.

- The annual Geography Action Week (usually held mid-October), organised by the Geographical Association, offers an ideal opportunity to put geography in the spotlight. For details of the current theme, and suggested activities, contact the GA.

Exploit the library as a useful resource for geography. Photo: Tim Price-Walker.

Conclusion

Different course ethos and market audience demands slightly different approaches to any marketing campaign. The key is to know your students or potential students and their parents in terms of marketing, at least meet their expectations and in most cases surpass them 'I didn't know that you could do that with geography ...'.

It is crucial that geography remains fresh and innovative. Teamwork in geography departments is essential for concept generation - new ideas should be tried and tested. With a little effort you can quickly convey your own enthusiasm for geography to students and parents alike. The returns are high - more enthusiastic students undertaking the subject as an option. If students and parents perceive you enjoying your subject, the knock-on effects will hopefully be many happy and worthwhile memories for all concerned!

Checklists for promoting geography

General

- [ ] Thematic/year group display work
- [ ] 'Geography in the News' or 'Geography making the Headlines' Noticeboard
- [ ] Laminate word-processed signs
- [ ] Signs locating the Geography Department, e.g. 'Discover Geography'
- [ ] Mobiles on ceiling in classroom
- [ ] Suspend plastic globes of the Earth
- [ ] Computers - set up a screen saver to promote the Department (e.g. Marquee on Windows)
- [ ] Networked software for whole-school use (See also Resource ideas for general promotion.)

Focus on year 9 and year 11

- [ ] **Year 9:** Lunchtime interviews when 'options' are being decided on. Noticeboard displaying information about the usefulness of geography
- [ ] **Year 11:** Lunchtime interviews for prospective A-level students. Noticeboard for school leavers and the range of jobs geography is appropriate for

Open day

- [ ] Room layout - create a spacious exhibition area
- [ ] Display work - laminate student's work for professional look
- [ ] Plastic globes
- [ ] Ensure computers are switched on and software ready for use
- [ ] Aerial photographs display of local area
- [ ] Examination papers with 'So you thought you were good at geography?'... then try some of these!' label
- [ ] A 'cover conscious' table to display student exercise books
- [ ] Video of a geographical programme, e.g. volcanoes (including music if possible) playing in room
- [ ] Working models on display, e.g. river simulation
- [ ] List of scheme of work on the board
- [ ] Maps of local area and magnifying glasses to provide interaction
- [ ] Geography students to act as 'guides'/demonstrators of CD-ROM equipment
- [ ] Fieldwork photographs and examples of work carried out on fieldwork
- [ ] Map out where parents come from using a local map and pins to create an instant sphere of influence display

Chapter 4: Promoting geography to your school's senior management team

Keith Grimwade

The introduction of the national curriculum in 1991 and the Dearing review in 1995 have had an impact on geography teaching in many schools (Grimwade and Thomas, 1996; Roberts, 1998; Donert and Grimwade, 1998). As part of a whole school curriculum review, the Head of Geography has often to make a case for geography to the school's senior management team, and sometimes to the staff as a whole. With the introduction of 'Curriculum 2000' it will be necessary to make this case again.

It is important to remember that many of your colleagues will have limited knowledge and understanding of contemporary school geography. I remember a colleague who, in all seriousness, said to me after invigilating an A-level decision-making examination, 'I didn't realise geography could be so interesting and useful'. We must also take into account the inevitable response of 'Well, they would say that their subject is important, wouldn't they'.

The implications of these two reactions are that we must ensure that everyone knows what teaching and learning in the modern geography classroom, *and in the field*, involves; and that we must present geography's *distinctive* contribution to the curriculum, as well as its wider contribution.

A small number of key sources will help you to assemble your arguments. Revisiting *A Case for Geography* (Bailey and Binns, 1987) is a good place to start. The International Charter on Geographical Education (*Teaching Geography*, 1995) provides many useful points. The Geographical Association's *Position Statement for Geography in the Curriculum* (GA, 1999) will also provide you with useful material. The articles referred to in the opening paragraph will allow you to substantiate much of what you say. The key points you should make are:

- Geography is the only subject which investigates real people in real places which is essential if our increasingly interconnected and interdependent world is to be understood, and if a respect for the diversity of cultures is to be fostered.

- Geography is the only subject which explores physical and human patterns and environments, e.g. the relationship between a river, its floodplain and settlement, is to be understood.

- Geography makes a distinctive contribution to our understanding of people–environment relationships, e.g. by exploring the socio-economic and political dimensions of atmospheric pollution in addition to its physical causes.

- Fieldwork in geography is distinctive because it investigates people in their environment and, as such, helps students to become aware of a diversity of environments and cultures, in their local area and beyond.

- Through geography, more than any other subject, young people learn how to use maps.

- Uniquely, geography supports the social and natural sciences. This is an essential contribution to the curriculum because it shows how gains in knowledge and understanding in the real world often require contributions from many disciplines.

- Geography makes a wider contribution to the curriculum by providing a context for work in literacy, numeracy and information and communication technology; by developing a range of relevant and practical skills such as problem solving and decision making; and by developing the qualities of good citizens and the attributes of spiritual, moral, social and cultural development.

Promoting Geography in Schools

Much has been said about geography's relationship with history, religious studies and social studies (see, for example, Walford, 1998). The reality is that geographers often find themselves within a Humanities Faculty, and very often in charge of it! Consequently, it is often necessary for the Head of Faculty to make a case for 'the Humanities' rather than for an individual subject, even though the subjects are taught separately. Steve Johnson provides some excellent advice and the following points have been adapted from his work (Johnson, 1996). These have been used successfully in promoting the humanities block on the 16-18 timetable in at least one school and adapt well for promoting our subject at all levels of the curriculum.

1. The humanities offer a distinctive contribution to the 16-18 curriculum through their common aims:

   - to encourage an understanding and respect for other people
   - to encourage a critical examination of a wide range of social, economic, political and environmental issues
   - to provide opportunities to interpret and respond to local, national and global events.

2. Humanities education deals with concepts which are vital to an understanding of today's world, e.g. freedom, responsibility, democracy, citizenship, equal opportunity, sustainable development.

3. Humanities education deals with key skills, e.g. empathy with people in different situations, recognition of one's own values and the influences on these, the recognition of the possibilities for future action.

4. Humanities education offers relevant, stimulating and interesting experiences, e.g. meeting and talking with people from different social, cultural, ethnic, age and gender groups, taking personal action on an environmental, cultural, social or economic issue.

5. Students need focused study to make sense of these aims, concepts, skills and experiences. Dropping humanities at 16 means dropping it at the very point when the majority of students are beginning to make sense of it all.

6. The humanities was dropped from the 16-18 curriculum in the Dearing review because the percentages did not add up, not because of any educational argument!

Another reality is that it will be far easier to promote geography to your senior management team if students are achieving well at all levels. This emphasises the point that promoting geography is about doing many things well and not just about high quality marketing and representational work. Although there can be little doubt that, with all the current pressure on the curriculum, the latter are needed more than ever before. But then so is geography - so be positive!

Promoting Geography in Schools

Chapter 5: A world of opportunities
Patrick Talbot with Alan Marvell

This chapter aims to help you to make your geography students aware of how they can use qualifications in the subject to add real direction to their future career planning. It also offers details and case studies of the type of careers paths they can choose.

Choosing to study geography at any level should be principally based upon an enjoyment of and an ability in the subject. However, when it is time to choose option subjects, students must give some thought to their possible future employment. Hence, careers-related materials which can support their decision to continue with geography - at either 14-16, 16-18 or as an undergraduate. While the focus of materials and advice within a school or college is naturally within the careers department you can help your geography students become aware of the contribution a geography qualification can make to their future career.

Photo: Sally Greenhill.

Promoting Geography in Schools

The CLCI classification

CLCI code	Job sector
B	Armed Forces
C	Administration, business & management
E	Art & design
F	Teaching & cultural activities
G	Entertainment & leisure
I	Catering & other services
J	Health & medical services
K	Social & related services
L	Law & related work
M	Security & protective services
N	Finance & related work
O	Buying, selling & related services
Q	Sciences, mathematics & related work
R	Engineering
S	Manufacturing industries
U	Construction & land services
W	Animals, plants & nature
Y	Transport

QOB	Chemistry
QOD	Biology
QOF	Physics
QOG	Mathematics
QOJ	Statistics
QOK	Economics
QOL	**Earth & environmental sciences** — Geography
QON	Food science
QOS	Materials science
QOT	Forensic science
QOX	Scientific laboratory work
QOZ	Others

Figure 1: The CLCI classification © Crown copyright. Reproduced by permission of COIC.

The COIC Signpost series

If you have a particular interest and ability in geography, the following list will show you some of the careers described in *Signposts* which have a connection to the subject.

- Air Traffic Control Officer/Assistant
- Airline Pilot
- Archaeologist
- Auctioneer/Estate Agent
- Cartographic Draughtsperson
- Coastguard
- Driver
- Driving Instructor
- Economist
- Engineer: Civil (Chartered)
- Farm Manager/Worker
- Forest Worker/Officer
- Geologist
- HM Forces: Commissioned and non-commissioned
- Landscape Architect
- Legal Executive
- Merchant Navy
- Meteorologist/Assistant Scientific Officer
- Motor Cycle Messenger
- Postman/woman
- Surveying Technician
- Surveyor: Agricultural/Land Agent
- Surveyor: Building
- Surveyor: General Practice
- Surveyor: Land/Hydrographic
- Surveyor: Mineral/Mining
- Surveyor: Planning/Development
- Teacher: Further/Higher Education
- Teacher: School
- Town Planner/Planning Technician
- Transport Manager
- Travel Agent/Clerk
- Travel Courier/Representative
- Warehouse person
- Zoo Keeper

Figure 2: The COIC Signpost series © Crown copyright. Reproduced by permission of COIC.

Where students can find geography-related careers materials

The two standard sources of material that a careers room should contain as a starting point for students are:

1. The Careers Library Classification Index (CLCI), which is an index of careers information. Any geography-related careers material should be found under QOL - Earth and Environmental Sciences (Figure 1).

2. The Careers and Occupational Information Centre (COIC) *Signposts* series is a set of information cards containing career suggestions that students can look up via subjects or skills. It lists those jobs which have a connection with geography (Figure 2).

3. The Geographical Association's new careers pack (Palôt, 1999).

However, it would be misleading to end any job search here. Students may ask what they can do with a geography qualification at 16+, 18+ or degree level. *Going Places* (Palôt, 1999), the GA's new careers pack has leaflets specifically aimed at answering these questions.

Explaining what geographers do

The answers to the question 'What sort of jobs do geographers go on to do?' are as varied as the question is simple. Geographers do, in fact, work in almost every field of employment. The Geographical Association's poster 'Where do all the geographers go?', an ideal source to illustrate this, is shown in Figure 3 (see also *Teaching Geography*, 1998; Palôt, 1999).

Photo: Wendy Garner.

Promoting Geography in Schools

Where do all geographers go?

Education, professional and social services

Fire/ambulance services
Law
Lecturing
Police service
Social administration
Social work
Teaching
Youth and community work

Environmental management

Architecture
Environmental health
Environmental research/consultancy
Estate/heritage management
Farming
Forestry
Groundsperson
Industrial development
Landscape architecture
National Trust
Nature conservation
Urban and rural planning

Information services

Archivist
Census officer
Central Office of Information
Film making
Information officer
Journalism
Libraries
Museums
Post office services
Publishing
Radio
Systems analyst
Telecommunications
Television

Figure 3: Where do all the geographers go?

Promoting Geography in Schools

the go?

Business and finance
Accounting
Advertising
Banking
Buying
Exporting
Insurance work
Market research
Marketing

Scientific services
Armed forces
Cartography (map making)
Coastguard
Computing
Geology
Hydrology and water services
Information and communications technology
Meteorology
Mining and quarrying
Oil companies
Photography
Research
Surveying

Management and administration
Building societies
Civil services
Estate agents
Freight distribution
Housing management
Local government
Opportunities abroad
Personnel management
Public relations
Retail management
Transport management

Leisure, travel and tourism
Air crew
Air traffic control
Courier
Hotel management
Recreational management
Sports management
Tour operator
Tourist boards
Travel agent

Promoting Geography in Schools

It is still possible, of course, to combine an interest in specific aspects of geography with employment. If your students enjoy hydrology, for instance, they could consider a job with one of the water companies or the Environment Agency. An interest in urban issues may suggest a career in planning; or interest in development could lead to work for a charity or aid agency. Wherever possible offer your students the opportunity to explore the relationship between the geographical skills they learn and those needed in certain jobs by working alongside, for example, local planners or architects. This could either be through visitors to the school explaining the nature of their jobs, via group visits to businesses (see photograph on page 41) or work placements for individual students.

To be fair, careers with a geographical flavour are not necessarily the preserve of the geographer. The maps we use in the classroom are likely to be produced by cartographers with a computing and mathematics background and the usual route into meteorology is via physics and mathematics, rather than geography.

In the case of geography-related GCSE/S-grades, GCSE Travel & Tourism courses, their students, for instance, will have studied the tourism industry and the impact that tourism has on various societies and economies. In 1998, 88 per cent of the 3633 candidates for GCSE Travel & Tourism gained an A*-G pass. These students are now in a strong position if they wish to further their studies in tourism. However, GCSE/S-grade Geography helps to develop the critical and analytical skills which can be applied to a tourism scenario. This means Travel & Tourism does not present an advantage at 14-16, merely a vocational aspect of study.

At Advanced level GNVQ, Leisure & Tourism students do have an edge over geography students when applying for employment in this vocational sphere. GNVQ Leisure & Tourism is essentially a business course dealing with the operational side, while in geography the nature of tourism is examined, often through a critical perspective. However, it is also true to say that many A-level/H-grade Geography students applying for Tourism-based courses and jobs and are well equipped to apply their geographical skills.

The most accessible source of quantifiable information on geography and careers is at the graduate level (Figure 4). Of the 6091 geography graduates from 1996, many remained loyal to the subject and went on to a teacher training course or to further academic study - MA, MSc or PhD in some aspect of the subject. Overall one-quarter of graduates in geography went on to some form of geography-related study or training. Three-fifths of geography graduates entered full-time employment. Most of those who graduated in geography in 1996 entered fields of employment that cannot be said to be primarily geographical: Sales, Marketing and buying, Financial work, and Administration and operational management. Nearly half of all graduate jobs advertised do not stipulate a specific degree and evidence suggests that a higher than average proportion of geographers get jobs in these three popular sectors. Geographers can, therefore, confidently compete for many jobs. This is borne out by the case studies of geography graduates which appear in the students' booklet in *Going Places* (Palôt, 1999). This booklet includes graduates views on how the subject has enhanced their careers. Many of those geography graduates appreciate the knowledge, skills and understanding which their degree has brought to their job.

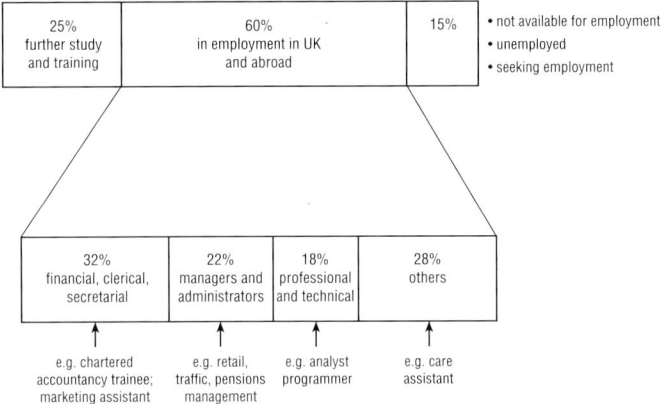

Figure 4: What did geography graduates do in 1996?

Sources used: HESA individualised student record 1995/96, reference July 1996. HESA first destinations supplement 1995/96, reference December 1996. © Higher Education Statistics Agency Limited 1997. Reproduced by permission of HESA. HESA cannot accept responsibility for any inferences or conclusions derived from the data by third parties.

Why geographers are so employable

The many skills contained in the courses geographers complete: *literacy, numeracy, graphicacy, computer literacy, analysis, individual research and group discussion,* make them very attractive to employers. Above all, geography's *synthesis of facts, figures, ideas and perspectives* provides ideal preparation for decision making at work (Figure 5). No wonder employers want geographers: the subject contains all the skills that employers deem most important (Figure 6).

The nature of peoples' working lives has changed with a 'job for life' now a thing of the past. As career paths become more varied, individuals will more readily transfer from working in one sector to another. The acquisition of transferable skills will be all-important: what better way to start than with geography?

Throwing out ideas

Geography needs to maintain its popularity in the top six GCSE/S-grades and A-level/H-grade subjects. (In 1997, it was the fifth most popular GCSE and A-level subject (in England) by recorded entry.) Intermittently addressing the sometimes overwhelming question of future careers can add credibility to the classroom approach. The contemporary relevance of geography makes it easy to suggest jobs which have a geographical flavour (Figure 7). Courses completed in secondary education are sequential, and career suggestions made to 11-14 year-olds may help crystallise choices made at 14 for GCSE/S-grade, at 16 for A-level/H-grade, at 18+ for a degree.

At 14-16 years of age students become eligible for work experience placements. Try to ensure that jobs of a geographical nature (see Figure 7) are used as a focus for work placements. You can also include work-related activites both inside and outside the

---

I have found over recent years that the Geographers we have recruited are well-organised, are able to structure their thoughts and actions most efficiently, and have very clear views of their preferred career paths.
        Pene Axtell, Graduate Recruitment & Training Manager, Carnaud Metalbox plc

... the Geographers who have been recruited this year were all strong in the area of information gathering.
        Jonathon Bond, Graduate Recruitment Manager, National Westminster Bank

We have taken Geography students in the past and I found Geography to be a good discipline from which to recruit our people. It gives them a broad appreciation of the earth's natural resources and an understanding of the properties of the materials with which we work.
        A. Ratcliffe, Training & Development Manager, Tarmac Quarry Products Ltd.

... we look for people with well developed communication and inter-personal skills and often find that degrees which allow for a variety of working situations (i.e. field trips, visits to foreign countries and placements working in teams and solo environments) provide us with a better quality of candidate than those courses which are much more 'research' based. On this basis, I am certain that Geography degrees fill these criteria.
        David Millbery, Personnel Executive, Newey & Eyre Ltd.

Geography graduates, particularly those who have experienced working in a team environment as leader and/or member, might well be able to demonstrate that they possess the qualities that we seek. The research and analytical skills that have been honed during their degree would be applicable in many commercial fields.
        Kevin Pimblott, Management Development Manager, Gallagher Ltd.

Figure 5: What recruitment managers say about geography.

Promoting Geography in Schools

What do employers want?

The skills sought by employers are ...
communication skills
  written
  oral
  technological
teamwork
flexibility
decision making
self management
powers of analysis
problem solving

What can geography students provide?

The skills that can be gained by doing a geography course ...

Communication skills
empathy with a wide range of viewpoints on issues has been encouraged

Written
a wide range of written work on the course: note taking, paragraph answers, essay work

Oral
classroom discussions on issues-related topics

Technological
a word processor was used to produce the individual study; a computer was used to search for information and to analyse it

Teamwork
fieldwork organised and carried out in groups

Flexibility
geographical study involves coping with a wide range of subjects, materials and scales

Decision making
some issues were examined via a 'planning enquiry' or decision-making exercise

Self management
an individual study had to be planned and carried out; investigations were set up in fieldwork; secondary data for the individual study was obtained from the library

Power of analysis
an analysis of data was carried out in the personal study and a wide range of techniques has been used in the course

Problem solving
information from a variety of sources was used to decide the outcome of an issue

Figure 6: Geographical skills in the job market: what employers want and what geography students will have.

Promoting Geography in Schools

Photo: Patrick Talbot.

Tectonic processes
Geologist, geophysicist, seismologist, vulcanologist

Geomorphological processes
Hydrologist, hydrographic surveyor, glaciologist, geomorphologist, geologist

Weather and climate
Meteorologist, television weather presenter, climatologist

Ecosystems and environment
Botanist, pedologist, estate manager, landscape architect, conservationist, forestry officer, countryside officer, National Park officer, working for English Nature, Scottish Natural Heritage, Countryside Council for Wales, Department of the Environment, the Countryside Commission, field studies officer, organiser of local schemes: Groundwork Trust, British Trust for Conservation Volunteers

Population
Demographer, census officer, researcher for a population agency, housing officer, social worker, immigration officer

Settlement
Town planner, civil engineer, land surveyor, town centre manager, site analyst for a major retailer

Economic activities
primary - farmer, forestry officer, mining surveyor
secondary - management in a manufacturing industry, cars, breweries
tertiary - recreation and leisure centre manager, tourism officer, working in the City, Civil Servant, traffic engineer, air traffic controller, transport planner

Development
Diplomatic service, British Council, Directorate of Overseas Survey, charity officer

Figure 7: Geography course content in relation to 50 specialist careers.

Promoting Geography in Schools

Present employment of ...

Richard	Geography teacher, school
Steven	Humanities co-ordinator, university
Jason	Geographic information systems manager
Matthew	Geographic information systems analyst
Simon	Planning development manager, working abroad
Andrew	Property fund manager
Dominic	Programme Funding Officer for a charity
Tom	Hydrocarbon logging engineer
Keith	Field operations manager, oil company
Barry	Transport operations manager
Nick	Landscape architect
Ian	Air traffic controller
Susan	Producer, BBC wildlife unit
Amanda	Geography teacher
Alison	Producer, BBC Radio 1 news
Janet	Environment planner
Dianne	Former teacher, now examinations officer
Sarah	Researcher - marine ecology
Elizabeth	Environment Agency worker
Susan	Marketing manager then retail manager
Julia	Information consultant, Bank of England
Philippa	Accounts management
Charlotte	Management - John Lewis Partnership
Clare	Human resources manager
Susie	Banking

Figure 8: Example survey of the careers of past students (of Hampton School (for Boys) and Bishop's Hatfield Girls' School).

Lesley

Geography, English, French at A-level
King Edward VI 6th Form College, Stourbridge
▼
Geography Degree
Birmingham University
▼
Personal Assistant's diploma
▼
Secretary to Personnel Director
▼
work in personnel
carpet manufacturer
▼
Personnel
Prudential Assurance
▼
Personnel Manager
Derwent Information

What Lesley says about her geography degree:
Apart from the course in Statistics, I could not say that much of the subject matter has had a direct bearing on a career in personnel. However, geography requires a fairly methodical and analytical approach and this is of benefit in the workplace.

Jason

Geography, Mathematics, Physics at A-level
Hampton School
▼
Geography Degree
Kingston University
▼
MSc in Geographic Information Systems (GIS)
University College, London
▼
12 month contract in GIS
Unisys
▼
GIS work
ER Mapper Ltd
▼
Management

What Jason says about his geography degree:
I have no regrets about taking geography. Most obviously, it helps me understand not only the processes behind the landscapes on satellite images but also the interactive nature of making considered business decisions.

Figure 9: Examples of career paths.

geography classroom which use specific geographical skills; for example, brief students to undertake a survey of a floodplain (see photograph) with a written report as a follow-up activity. Students should be given copies of their record of achievement and, you could consider giving them copies of the post-16 and post-18 skills checklists to complete (see Palôt, 1999). All of these activities will help students prepare for working life and some will prepare them for college/university/job interviews.

A survey of past students who studied geography and their subsequent studies or careers paths can provide a useful discussion point (Figure 8). The use of case studies is another simple way to emphasise the relevance of geography to a variety of careers (Figure 9). These can be compiled from a skeleton of background information together with some annotation which can point out common or significant trends. Drawing on the experience of past students can personalise the use of role models. They may be prepared to come back into school, for a careers convention or an informal visit to the geography department. Ask them to comment on how geography has helped in their career (you could ask them to complete a post-18 questionnaire (see Palôt, 1999) and keep it on file).

Content and skills

With employment in mind, geography recommends itself as an option in both content and skills. Any geography course will suggest many relevant employment areas and will teach students a wide range of skills that are vital in employment. Outside teaching and lecturing, few individuals may call themselves professional geographers; however, those with geography qualifications readily recognise the benefits it has given them for the world of work.

Promoting Geography in Schools

Chapter 6: Extra curricular activities in geography clubs and societies

Tony Dodsworth

This chapter outlines how the setting up of a geography club or society in your school can help you promote the subject to your students and their parents.

How many times have you devised activities that your students would enjoy and which would develop their knowledge and understanding of the world but had to shelve due to lack of time or because they did not seem to match the appropriate programme of study? Setting up a geography club or society will allow your department to convince students that 'Geography is fun' (see Dodsworth, 1995). Learning is not the prime objective - the geography club should enable students to enjoy activities that have geography at their heart. A geography club should broaden students' perceptions of the subject and develop their interest in the world - not just through classwork and homework, but through activities that are not considered 'work' at all.

The practicals

To set up a geography club or society an initial outlay of £250-£400 will be required. If your school has a departmental development fund, apply for part of this. Make the senior management in the school aware of the social benefits of taking students 'off the corridors' at lunchtime while involving them in worthwhile activities. You can also add that the resources purchased for the club will almost certainly be of use within the classroom. A small annual fee can be charged for membership of the club, but you may decide that this is not necessary. If money is raised it could be used to buy more resources for the club or donated to charities with a clear link to geography, such as Shelter, Intermediate Technology or a rainforest charity like Equafor.

Setting up

To launch a geography club first choose an appropriate name, ours is called 'Globetrotters' and create a cartoon character which should appear as a 'logo' on all the club's printed material. Produce an advertising sheet (Figure 1), an information sheet (Figure 2), application forms and club cards beforehand. To avoid being swamped it is best to start with one introductory lunchtime session for each targeted class. The aim of this session is to make the students aware of what is on offer in the geography club. Once students decide they wish to join, ask them to complete an application slip and supply each one with a club card. Funding at present comes from departmental capitation although CD-ROM material has been bought following a special request to a Special Development Fund in the school.

In most cases it will be best to run the geography club at lunchtime - perhaps once or twice a week - and initially to target students in years 7 or 8. Use a 'geography room' if possible (easier to set up and put away 'equipment') and devise ways of developing the club so that over time older students can become involved in activities. With increasing use of CD-ROMs it can be useful to book a particular lunchtime each week in a computer room equipped with a number of computers that can run CD-ROMs.

Figure 1: Example advertising leaflet.

Do you want to join the globetrotter club?

- Do you enjoy playing with electronic games, doing jigsaws, playing board games, doing puzzles and quizzes and using maps and atlases?

- Do you find the time at lunch break sometimes drags a bit?

- Do you want to become a real 'wiz' at knowing where places are in the world?

- Do you want to get a club card that gets your achievements stamped on it and qualifies you for three different certificates of merit?

Yes? Then the Globetrotter Club is for you.

The Globetrotter Club is being set up by the school's Geography Department to provide for you a fun way to find out about the world by using GeoSafari game machines, map jigsaws, board games like Journey Round Europe and wordsearches and puzzles about places. You will be able to amaze everyone at home with your knowledge of the world and prove how good you are by getting your club card stamped and receiving certificates of merit.

The Globetrotter Club will be based in Room 5 and meetings will be held on Tuesdays from 12.10 to 12.45. There will be an introductory session for each year 7 class so you can find out what is involved before you commit yourself to joining by filling in an application form and receiving your official club card. If the demand is there it may be possible to run the club for another lunchtime session each week. There is also the possibility of selling a club badge and other items with a global theme, such as pens, stickers and pencil-sharpeners.

Keep your ears open for details about your class's introductory session and come on in!

Figure 2: Globetrotter Club information leaflet.

Promoting Geography in Schools

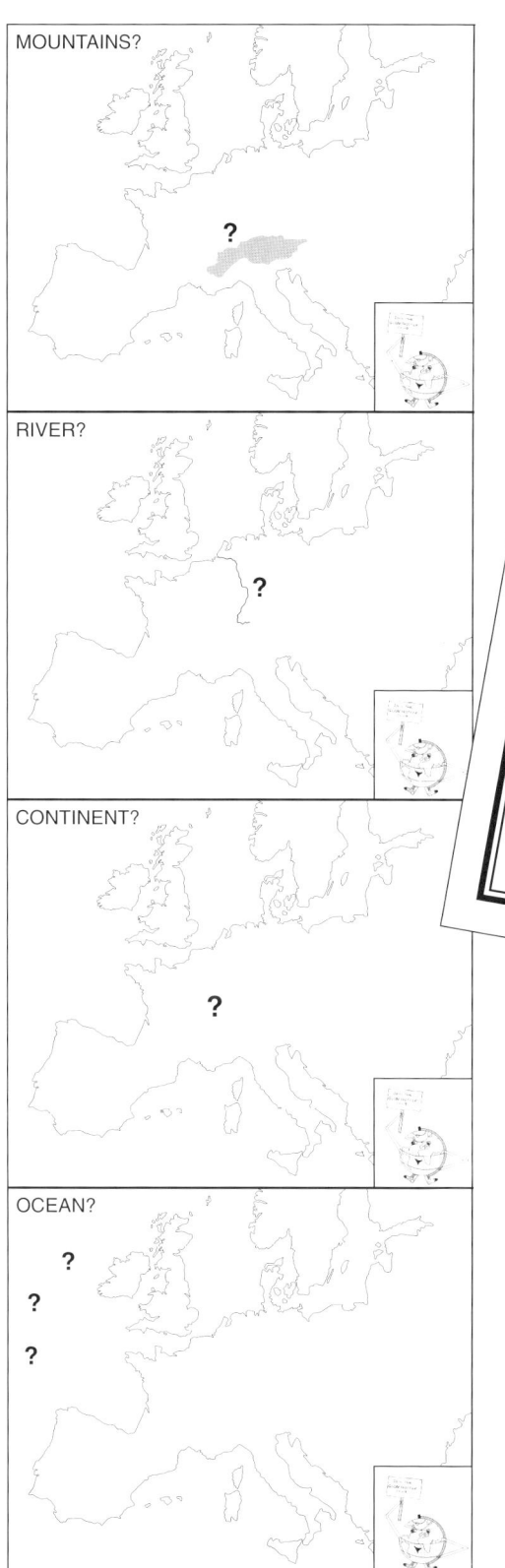

Figure 3: Quiz map cards for introductory sessions.

Figure 4: Example certificate for specific club activity.

Club activities

With their club cards students can volunteer for short tests on their place knowledge centred on the maps in the National Curriculum. Produce a series of quiz map cards (Figure 3) and conduct oral tests of a 'fun Mastermind' type. Students who score well at 'Mastermind' earn stamps on their cards which build up to certificates (see Figure 4), these can be signed by the Headteacher and presented in year assemblies. The activities in the club can be geared towards helping students prepare for the tests by reinforcing their place knowledge. However, only students who volunteer for the tests need be involved; others will enjoy club activities without being tested.

Promoting Geography in Schools

Key resource - GeoSafari machines
Can usually be found in large toy shops (such as Toys 'R' Us', but compare prices with the Argos catalogue and Children's World. Set of map cards free with every machine. Cost £50 approx. The set of cards on Ecology may be of interest (£9.99). Producer: Educational Insights Ltd., Unit 8 Fulton Close, Argyle Way, Stevenage, Hertfordshire SG2 1AF.

Games
Explore Europe and *Race Around Britain*. Both of these are produced by Ravensburger Ltd (Bessemer Close, Bicester, Oxfordshire OX6 0JD). Again look in major toy shops and department stores, e.g. John Lewis.

Other games involving maps: *The London Game* (Toybroken Ltd, Huntingdon, Cambridgeshire) and *Scotland Yard* (Ravensburger). Magnetic, pocketable: *Travel World* (SYU Creation, Watton, Peterborough PE4 6HD).

Jigsaws
Look in specialist jigsaw shops or major toy shops and department stores. Where possible choose strong cardboard jigsaws, flimsy card ones will simply disintegrate. A number of picture map puzzles are available, for example: British Isles/Europe/USA/Australia/etc. - 500 pieces, suitable for age ten plus (J.R. Jigsaws, Handley Printers Ltd., Stockport SK6 2BR).
World Map - 1000 pieces (F.X. Schmid, D-83209 Prien - made in Germany).
British Isles - 250 pieces (Usborne Publishing Ltd., 83-85 Saffron Hill, London EC1N 8RT).

Books
Look in good bookshops and department stores.
The Usborne Geography Quiz Book by Marit Claridge and Paul Dowswell. Usborne Publishing, 1992.
The Usborne Book of Earth Facts by Lynn Bresler. Usborne, 1986.
Project Geography by Philip Boys. Puffin Books, 1995.
Worldwise Quiz books: 5 , (1991), *6* (1992) and *7* (1999) Geographical Association.
Know the World by David Green. Tarquin Publications, 1993. ISBN 0 906212 90 1.

World Puzzle Book by Andrew Dalwood. Geographical Association, 1997 (revised edition). ISBN 1 899085 38 6.
The Magic Globe by Heather Maisner. Walker Brothers, 1995. ISBN 0 7445 4384 3.
Map and Maze Puzzles (Advanced) by Sarah Dixon and Radhi Parekh. Usborne Publishing Ltd., 1993. ISBN 0 7460 1579 8.

Information Technology materials
Where access to a computer is possible you can use any floppy disk based 'atlas' material such as *MacGlobe*. Sherston Software. (Tel: 01666 840433) produce a mapwork game on CD-ROM called *Map Detective*, which is very good, plus interesting OS map exercises on disks entitled *Maps 1* and *Maps 2*.

There are increasing numbers of appropriate CD-ROMs which can be used in a geography club. Among these it may be worth having a look at *Where in the World is Carmen Sandiego?* (Broderbund), *Geobee Challenge* (National Geographic Society), *Material World* (Anglia) and *Discover London* (Ordnance Survey); all of which are available from either REM (Tel: 01458 253636) or AVP (Tel: 01291 625439). If you do not have access to a CD-ROM in or close to your department mount a campaign to obtain one. Stress to the fundholders the importance of computer software as a visual source and the information retrieval skills required in geography and ICT.

Other materials
Keep an eye out for other puzzles, the *Rubic Globe* which British Home Stores sell around Christmas time would be useful, as would wood block puzzles showing a world map. Look in magazines for puzzles involving 'geographical matters' and check the GA resources catalogue and *Teaching Geography* and *Primary Geographer* which often contain game boards/ideas for games.

Make up 'flash cards' with parts of the world map or ask the students to devise games and hold competitions and quizzes.

Award certificates or prizes for competence in map recognition, etc.

Figure 5: Materials for a school geography club.

Promoting Geography in Schools

The types of materials and activities that might be considered for a geography club are illustrated in Figure 5. The *GeoSafari* machines usually prove the most popular of the resources available; many students will have seen them advertised on children's television. A set of more than 30 map cards is usually included in the price of a *GeoSafari* machine and a separate set of cards on ecology involves geographical information. You or your students can devise extra map cards, e.g. based on your school area. Consider buying mains adapters for your *GeoSafari* machines - this saves constantly having to replace batteries. Competitions based around the *GeoSafari* machines are very popular, individuals or class teams can compete against each other, scoring can involve total scores and time taken.

Obtaining club access to a computer allows software or CD-ROM packages with geographical content to be purchased for use within the club. Computer packages often present information visually and as geography relies to some extent on this type of presentation, a strong case for the central role that geography plays within a school in developing students' visual skills can be made. Link this to the importance of enquiry skills within geography alongside the needs of ICT in the national curriculum and you have a very good case for persuading the school's senior management to locate a CD-ROM either in or with very good access for the department! Students find using computers and CD-ROMs of great interest and this further motivates them to discover more about the world in which they live. An 'atlas' program, such as *MacGlobe*, can be linked to numbered Task Cards (see Figure 6). Tasks should be graduated in difficulty, as students work through the set they must then use increasingly complex information retrieval skills to complete the tasks. A record sheet (Figure 6) helps students keep a log of the tasks they have completed. Geographical CD-ROMs can be utilised in the same way.

The Land Use-UK Survey conducted in the summer of 1996 provided an ideal opportunity to involve students in geographical activities outside the usual format. Saturday trips for geography club members can include a visit to a sea-life centre, zoo or butterfly house (see photograph) together with a short walk in the countryside. Stage 'one-off' events at lunchtime - such as a treasure hunt around the school buildings and school grounds using map-

Students enjoying a Saturday trip to Tropical World, Roundhay Park, Leeds. Photo: Tony Dodsworth.

Globetrotters taking part in a 'Rainforest feast'. Photo: Tony Dodsworth.

MAC GLOBE

Task 5
Use the country maps

Find the map of Argentina

a) What is the population of Argentina?

b) Draw and colour in the flag of Argentina.

c) Briefly play the National Anthem.

d) Name the capital of Argentina.

e) Apart from the capital, name four other cities in Argentina with a population of over 500,000.

f) Name the longest river in Argentina.

g) Which mountains are found in the west of Argentina?

h) Name the island off the southern tip of Argentina.

MAC GLOBE

Task Cards

Name .... Class ....
Tick and date the boxes below when you have completed a task.
Do the tasks in order.

	Date		Date
1		11	
2		12	
3		13	
4		14	
5		15	
6		16	
7		17	
8		18	
9		19	
10		20	

Figure 6: (left) sample MacGlobe task card and (right) record sheet for completed tasks.

reading skills plus cryptic clues. Hold a 'Rainforest feast'. Ask students to prepare (preferably cold) dishes at home but specify that the essential ingredients must originally come from rainforest 'products', e.g. chocolate, coconut (see photograph). Have the entries judged by a food technology specialist as on the television programme *Master Chef* and award a prize. Slice up all the entries into small pieces for sampling alongside small portions of exotic fruits like mangoes, lychees, etc. This enables students to grasp just what originated in a tropical rainforest environment. Poster competitions based on a world theme such as that illustrated in Figure 7 can also be very successful.

Produce a termly newsletter giving information about club activities and include reports on any club trips or events. It need only be two sides of A4, but could be expanded to four sides (see for example Elstone, 1997). Commission different groups to write sections on general or specific activities. Divide the newsletter into sections such as: 'Fun facts' and a quiz. If at all possible include scanned in or pasted on photographs. These make the overall appearance much more attractive.

One other attraction of establishing a club is that the activities involved can be easily 'shown off' at open evenings and options evenings. 'Active' displays and the variety of club activities never fail to impress parents and school governors. In addition, anything that opens students' eyes to the possibilities that exist within geography or increases motivation and enjoyment while discovering more about the world, must be encouraged. This is what a school geography club or society will help you to achieve.

Figure 7: Student instructions for a club poster competition.

Globetrotter's Poster Competition

Your assignment is to produce a poster that brings home to people looking at it that the environment of the world is under threat.

The particular theme we are going to use for each poster is that somewhere within its design there should be included a picture of a globe. Above you can see three examples of what is needed but remember you could include a battered globe covered in tanks, globes overcrowded with people or anything else that you think would get across the message.

Make your poster as colourful as possible and include on it a short slogan.

Promoting Geography in Schools

"I KNOW, WE'LL TAKE THE WHOLE YEAR OUT TO LOOK AT ENVIRONMENTAL ISSUES!"

Chapter 7: Showing some initiative: a case study

Huw Jackson

This chapter provides a case study of one school's geography department. It concentrates on the efforts of the geography teachers in recycling previous ideas and introducing new initiatives in order to encourage more students to study geography.

The school

Bexleyheath School is a 1700 student mixed local education authority comprehensive in south-east London. The school draws its students from the Thames-side industrial areas and middle income pre-war suburban estates. Selection at age 11 is a major influence on transfer to secondary education in the Borough of Bexley which means that, in theory, the most academically able are creamed off by selective schools. It could therefore be suggested that the school is in the mould of the old secondary modern. In fact, the whole ability range is represented with 40% of candidates in 1996 gaining grades A-C at GCSE, with A-level an important component in post-16 provision.

The geography department under pressure

Geography has historically been a strong department in the school with a good take-up at GCSE and A-level. There are four full-time and two part-time staff, all with a strong commitment to the subject. Two staff have responsibilities as Year Heads. More recently, due to national initiatives and changes within the school, major body blows have been inflicted on the department. Nationally, history and geography are now optional subjects at key stage 4. This lead to both subjects becoming Free Options at Bexleyheath. It is also in competition with subjects which (in 1996) included an additional language, single science and drama. Numbers taking geography immediately fell from around 150 (half of the year 9 cohort) to under 100 (the 1996 take up in year 10 was 87). There is also strong competition from local colleges for post-16 education which seriously affects the number and quality of students staying on at Bexleyheath. However, the introduction of GNVQ courses and a heavy investment in a range of courses, staff and time has helped to counter this trend. Numbers in each year group have averaged 6 or 7 in the last few years and this, coupled with financial constraints, has led to the amalgamation of years 12 and 13 into one teaching group.

Internally, decisions led to geography being reduced from five to four hours at key stage 3, with year 9 being crucially reduced to just one hour a week. Apart from restricting opportunities to design a broad and relevant syllabus, there are now constraints on our ability to 'sell' the subject. This has been further compounded in year 9 by the school's decision to offer vocational subjects as options in competition with geography, to provide for less academically inclined students.

Time to take stock

It was only a few years ago that we were enjoying the fruits of imposed expansion and taking on additional staff. However, our school now has too many staff, so all six geography staff devote 25% of their time to other subject areas: mathematics, English and religious education. Declining morale was further eroded by the decision to split the Geography Department between two sites, imposing daily 400m hikes.

It became clear that our former success was partly due to external decisions - we did not really have to 'sell' geography. The straight choice between history and geography led to a comfortable take-up and a large pool of potential A-level geography students. The relationship between staff and student was sufficient to sustain us. The developments outlined above as well as the potential for future change made it clear that we had to act if we were to ensure that geography was to survive as a distinct subject.

Promoting Geography in Schools

Selling geography

Selling geography was something we had done for many years in a rather mechanical way. At open evenings we put on a bit of a show aimed at parents. This essentially says that geography is a nice interesting subject which gains respectable results. The fact that the Head of the school expressed satisfaction was appropriate reward for our endeavours. For students, we produce single A4 leaflets outlining GCSE and A/AS-level courses. Our corridor displays attracted praise for their colour and success in improving the school environment.

The pressure on the subject made us think more precisely about how we attract students to geography and how, whatever methods we chose, were we going to satisfy our own aspirations for it. The aim is to sell geography as a product and promote it through raising its status above other subjects. This would involve teachers and students as well as parents, school management and governors.

Fieldwork was a weak point of the department. This was mainly due to the priorities of the previous management. However, the more supportive attitude of the new Head enabled us to offer a residential experience for A-level students. This took place in South Wales for the first time in 1996 with a similar year 10 trip in 1997 to Dorset. We also provide day trips for key stage 3. Fieldwork is clearly a tool for promoting geography that sets it apart from other subjects. We also offer extra-curricular activities aimed at specific year groups, which can be linked to studies in the classroom, a conservation day (see Figure 1) can be linked to our students' classroom studies on the environment. We take every opportunity to publicise such events with corridor displays and articles in the weekly staff bulletin.

Course provision at key stage 4 also came under scrutiny. With an eye on the potential impact of proposed vocational courses, we are now developing the WJEC Certificate of Educational Achievement. This represents a welcome opportunity to cater for lower ability students by allowing them to succeed at different levels. The Welsh Joint Education Committee (WJEC) Certificate could be used as a promotional tool to attract candidates to geography. With option choices in mind in year 9, we deliberately targeted girls. At key stage 3 girls are

Figure 1: Geography Department year 9 information leaflet.

Figure 2: Geography Club field trip poster.

often more able in geography than boys. The girls then opt out of geography at key stage 4. To encourage them to study geography beyond year 11, we target potential A-level candidates far more methodically than before. Personal encouragement

Promoting Geography in Schools

Geography Club members enjoying the visit to Drusillas Zoo, near Eastbourne. Photos: Huw Jackson.

works best, certainly at GCSE results were clearly biased against girls in the past. In addition we have revamped our corridor displays with the same aims in mind. Thus we provide year 11 with displays of sixth form fieldwork early in the year.

So what's new?

So far you may argue that we are doing things which any self-respecting department would already be doing, that we were lulled into a false sense of security by our previous success. To some degree this is a true, but discussion of this issue would not serve any purpose in the context of this chapter. What became clear as we implemented these initiatives into action was the need to make an impact on students much earlier in their secondary school life, in years 7 and 8.

In Bexleyheath, as elsewhere, the bedrock of our work is an enthusiastic team of teachers. To complement their work and enhance the status of geography we developed several new initiatives.

In year 7 we run a Geography Club (see also Chapter 4). The *GeoSafari* computer game helps to maintain a regular clientele. Saturday trips, where all the emphasis is on learning through an enjoyable day out (Figure 2), has proved to be a winner with students and parents alike. We combined a visit to Brighton Sea Life Centre with a walk on the South Downs at Devil's Dyke. It was obvious that these were new experiences for many students and was in some cases reminiscent of a pit pony's rare experience of fresh air and green grass! A later Geography Club trip included a visit to Drusillas Zoo in Eastbourne (see photographs). While the trip had implications for effective control outside the classroom environment, it was great to see students soaking in their surroundings. Our Geography Club is also developing other ideas, which include producing a mural on geography and sponsored events.

It is difficult to maintain a club after school hours: competition comes from other attractions, a 45-minute lunch break, the school calendar's first call on staff for meetings and winter darkness all conspire to produce problems of continuity and maintenance of interest. The teacher's persevere because the Geography Club brings its own rewards.

Another feature in the department is a termly newsletter called *Geogers* (Figure 3). Now in its fourth year, this is easy to produce using *Pressworks* on my PC at home or on *Publisher* in school. The actual articles write themselves. Two A4 sides limits what is possible, so teachers must be creative with space. It is valuable to include contributions from students, for example, following up the Saturday trips in Club sessions has produced interesting stories and poetry. The aim of the newsletter is to inform, to recognise student achievement and to

Promoting Geography in Schools

promote the subject. It includes reports of the Geography Club, Revision Club, field trips, a quiz and the names of students who have achieved something notable in class (e.g. the best group project). The newsletter has also proved useful on Open Evenings as a handout which gains a positive response.

A third initiative has been to award Certificates of Merit (Figure 4). Bexleyheath School classes are streamed which made us decide to award Certificates across the ability range. Thus at least one student and not more than two from each class is awarded a 'Certificate for outstanding effort in geography'. The limit maintains an exclusive air to the Certificates and separates their award from existing school reward systems. In practice, it allows the head of geography department to get his face known whether the award is made in the class or in an assembly. More importantly students' achievements are recognised.

While each of the above activities enjoy varying degrees of success, collectively they provide opportunities for publicising geography at all levels within the school and beyond to parents. This creates the feeling that geography is always doing something and that the geography staff are doing things for students.

Bums on seats?

The final success or otherwise of what has been additional work can only be measured by the numbers who actually opt for geography. I could quote figures, but they would not be significant. I'm happy to report that geography is holding its own as one of the most popular options while history is struggling. Our numbers at key stage 4 remain around the 90 mark, with an increased proportion of more able girls and less able boys. Any future review of our department should focus on which activities were the most effective in attracting and retaining students to geography.

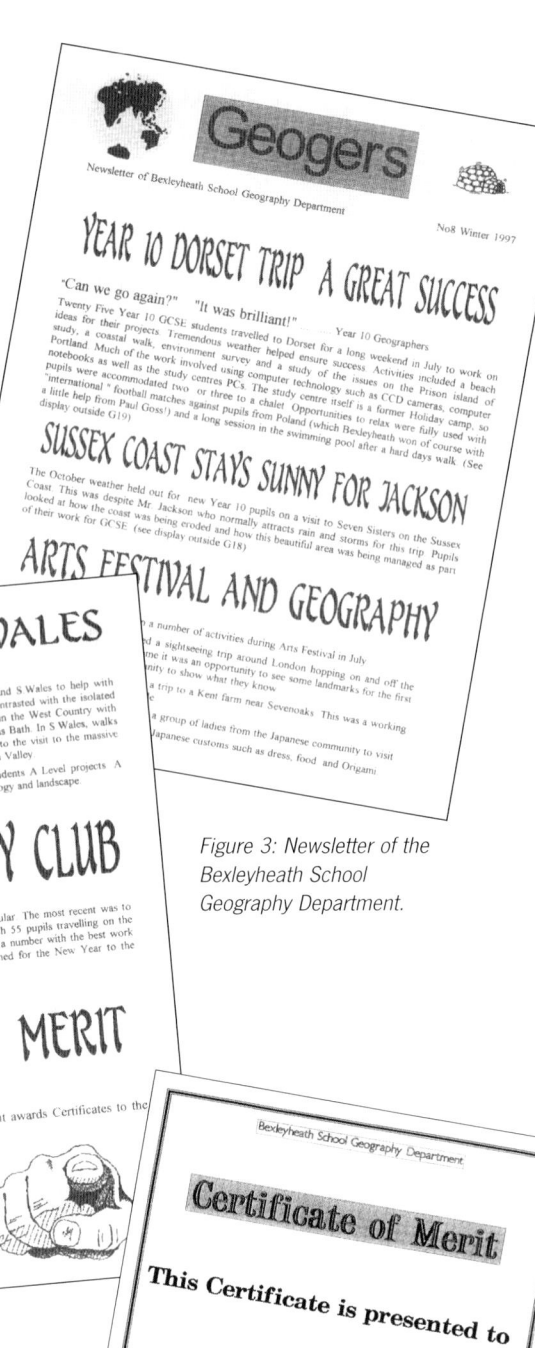

Figure 3: Newsletter of the Bexleyheath School Geography Department.

Figure 4: Certificate of Merit for outstanding effort in geography.

Bibliography

Bailey, P. and Binns, T. (eds) (1987) *A Case for Geography*. Sheffield: Geographical Association.

Barrett, H. (1996) 'Education without prejudice' in Bailey, P. and Fox, P. (eds) *Geography Teacher's Handbook*. Sheffield: Geographical Association, pp. 29-37.

Cawley, R. (1997) 'Display - the forgotten teaching method?', *Teaching Geography*, 22, 1, pp. 26-9.

Conolly, G. (1997) *Rediscover Geography*. The Geography Teachers Association of New South Wales, Australia.

Dalwood, A. (1997) *World Puzzle Book*. Sheffield: Geographical Association.

Dodsworth, T. (1995) 'Setting up a school geography club', *Teaching Geography*, 20, 4, pp. 185-6

Donert, K. and Grimwade, K. (1998) 'The state of geography in secondary schools: two years on', *Teaching Geography*, 23, 2, pp. 67-70.

Dove, J. and Tinney, S. (1992) 'Using classroom display as a record of achievement', *Teaching Geography*, 17, 2, pp. 57-60.

Durbin, C. (1995) 'Using televisual resources in geography', *Teaching Geography*, 20, 3, pp. 118-21.

Elstone, D. (1997) 'Producing a school geography magazine', *Teaching Geography*, 22, 1, pp. 35-7.

GA (1995) International Charter on Geographical Education, *Teaching Geography*, 20, 2, pp 95-9.

GA (1998) 'Choose geography: a leaflet and poster for students and teachers' (special pullout section), *Teaching Geography*, 23, 2, pp. 80-4.

GA (1999) *Position Statement on Geography in the Curriculum* (leaflet). Sheffield: Geographical Association.

Grimwade, K. and Thomas, T. (1996) 'Geography in the secondary schools: a survey', *Teaching Geography*, 21, 2, pp. 37-9.

Jackson, H. (1996) 'Promoting geography', *Teaching Geography*, 21, 4, p. 140.

Johnson, S., (1996) 'Towards 2000: a humanities entitlement at key stage 4', *Teaching Geography*, 21, 2, pp. 96-7.

Kent, A. (ed) (1990) *Selling Geography*. Sheffield: Geographical Association.

Kent, A. (1996) 'A place in the heart', *Times Educational Supplement*, 18 October.

Kent, A. (1997) 'Challenging geography: a personal view', *Geography*, 82, 4, pp. 293-303.

Kent, A. (forthcoming) 'Image and reality', A research forum, *International Journal of Research on Geographical and Environmental Education*.

Marvell, A. and Smyth, T. (1996) 'Should you consider GCSE Travel & Tourism?', *Teaching Geography*, 21, 2, pp. 92-4.

Nagle, G. (1996) 'Crime in Oxford', *Geofile*, article no. 287.

Nagle, G. (1997) 'Agriculture issues in the UK', *GeoActive*, article no. 160.

Nagle, G. (1997) 'The geography of disease; UK focus', *GeoActive*, article no. 163.

Nagle, G. (1998) *Geography Through Diagrams*. Oxford: Oxford University Press.

Nagle, G. and Spencer, K. (1996) *A Geography of the European Union: A regional and economic perspective*. Oxford: Oxford University Press.

Nagle, G. and Spencer, K. (1996) *Advanced Geography: Revision handbook*. Oxford: Oxford University Press.

Palôt, I. (1999) *Going Places: A geography careers resource pack*. Sheffield: Geographical Association.

Price-Walker, T. (1995) 'Open day success!', *Teaching Geography*, 20, 3, pp. 133-4.

Qualifications and Curriculum Authority (1998) *Geography and History in the 14-19 Curriculum* (leaflet). London: QCA.

RGS (with IBG)/GA (1998) *Geography: an essential contribution to education for life* (leaflet). London/Sheffield: RGS (with IBG)/Geographical Association.

Roberts, M. (1998) 'The impact and legacy of the 1991 Geography National Curriculum at key stage 3', *Geography*, 83, 1, pp. 15-27.

Rogers, S. (1998) 'Presentation software', *Teaching Geography*, 23, 3, pp. 150-1.

Scottish Association of Geography Teachers (1995) *What is Geography* (a video).

Speake, J. and Donert, K. (1998) 'Approaches to studying the geography of crime', *Teaching Geography*, 23, 1, pp. 5-9

Stanfield, J. (1995) 'Topicality in geography teaching', *Teaching Geography*, 20, 4, pp. 168-70.

Swift, D. (1996) 'Is there geography for all at key stage 4?', *Teaching Geography*, 21, 2, pp. 83-6.

Walford, R. (1998) 'Geography: the way forward', *Teaching Geography*, 23, 2, pp. 61-4.

Worldwise Quiz Book no. 5 (1991), *Book no. 6* (1992) and *Book no. 7* (1999) Sheffield: Geographical Association.

Organisations and addresses

Note: all website addresses should be prefixed by: http://www.

BTEC (now part of Edexcel), Stewart House, 32 Russell Square, London, WC1B 5DN. Tel: 0171 393 4500; website: edexcel.org.uk

City and Guilds of London Institute (C&G), 1 Giltspur Street, London EC1A 9DD. Tel: 0171 294 2468; Fax: 0171 294 2405.

The Geographical Association, 160 Solly Street, Sheffield S1 4BF. Tel: 0114 296 0088; Fax: 0114 296 7176; website: geography.org.uk

Qualifications and Curriculum Authority, 29 Bolton Street, London W1Y 7PD. Tel: 0171 509 5555.

Royal Geographical Society, 1 Kensington Gore, London S7 2AR. Tel: 0171 589 5466; website: rgs.org.uk

Royal Scottish Geographical Society, Graham Hills Building, 40 George Street, Glasgow G1 1QE. Tel: 0141 552 3330; website: RSGS@strath.ac.uk

Royal Society for the Encouragement of Arts, Manufacture and Commerce (RSA), 6-8 John Adam Street, London WC2N 6EZ. Tel: 0171 930 5115; website: rsa.org.uk

Scottish Qualifications Authority (SQA), Hanover House, 24 Douglas Street, Glasgow G2 7NQ. Tel: 0141 248 7900

The Travel & Tourism Programme, 3 Redman Court, Bell Street, Princes Risborough, Bucks HP27 0AA. Tel: 01844 344208; Fax: 01844 274340.